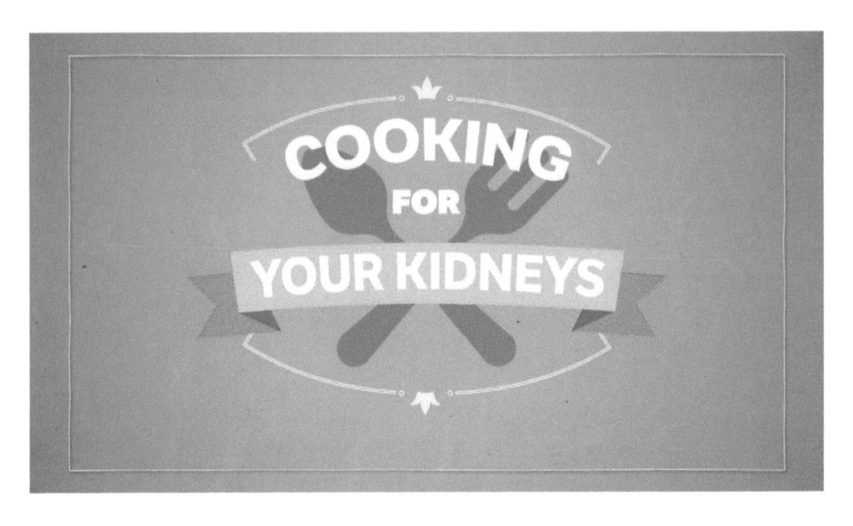

Recipes, Nutrition, Food Science, and Insights from a Chef and Chronic Kidney Disease patient.

John A. Vito

Published by JVFS, Inc, Rochester, NY

Library of Congress Control Number: 2019902656

Vito, John A

ISBN: 978-1-7338094- 0-5

Design by JVFS, Inc.

Photos: Maria Spinelli - Cover photo, pages 24, 30, 49, 112. back cover video shoot
Jon Feldman - page 85,138. back cover photo
John Vito - all others

To my mother and father

Acknowledgements

It took a great deal of effort to pull myself out from under the rubble after my diagnosis, and to finally arrive at a place where I could write this book.

My own efforts were never going to be enough and without solicitation, many came to my aid. They provided the necessary force to not only get this project started, but to see it to completion.

My friends Jon and Jill pushed me to begin when I was starting my new life and trying to find my way. They offered constant positive energy, and a strong belief that is often lacking during the process of chronic illness.

Many others wanted to help when I became ill and my friends Margaret and Joe provided a great deal of assistance in every area. They organized the efforts of others and acted as my points of contact for the many offers so that I was not overwhelmed while navigating my new existence. They made certain that I never felt completely exiled from my former life, and treated me much the same as they did prior to my illness. They were, and still are, a constant source of support in all areas and through the many years of my ongoing illness.

Tom and Sharon took me into their home and reworked their lives to accommodate my needs during transplant surgery and for the 6 weeks of recovery that followed. A big ask for anyone that they both seamlessly embraced.

Franlee taught me the book business from the publishing side and assisted with the process from the early stages. I have a long to way go, but she provided me the direction to apply my efforts.

I am grateful for the constant support in what has been a long process in reclaiming some of that which has been lost to chronic kidney disease.

Table of Contents

Introduction

I have chronic kidney disease, or CKD as it is shortened to in the medical and care community. It means my kidney function was deteriorating over time and they were not ridding my body of the toxins and fluid that accumulate. However, I was not aware of it, not until I awoke in the ICU connected to a dialysis machine.

My kidneys had shut down and my blood pressure was two hundred and something over one hundred and something. I was not aware of that fact either. Blood pressure is called the silent killer for a reason, a reason I am much more familiar with now.

The surroundings suggested that my current situation was serious. Finding out family and friends had gathered outside the ICU, waiting to see if I was going to wake up, pointed out it was more than a suggestion. I glanced to my right and saw a woman sitting and staring at that rectangular machine with great intensity. She did not look at me at all, making me wonder about her thoughts on my condition.

I was groggy, a bit delirious, and certainly confused. A few family members were allowed to come into the room for a short visit, two at time. The staff only allows family to enter the ICU and I heard a bit of a tussle outside my door. I could see two men in suits walk through telling the doctor when asked, “Of course we are family.” She looked at me and I nodded as they both walked right past her.

I am not sure if it was my grogginess up to that point or the others ability to hide what they were seeing, but the expression on the faces of my two friends confirmed how bad my condition actually was. The look of shock that quickly transitioned into fear got my attention, and I started to further investigate my surroundings. I had two tubes coming out my chest connected to the machine. They were both filled with blood, as was the right side of my chest and the sheets underneath. I learned later that getting those tubes in was not an easy task. It all was coming into focus.

Sometimes kidneys shut down for a short period of time and will start functioning again. This is called acute kidney failure and recovery is possible. Chronic kidney failure damages the kidneys over time and they do not.

Over the next several days and multiple tests, it was confirmed that my kidney issues were chronic. The limited function that I currently had was all that I was going to get out of them. In addition, the levels were so low that dialysis became a daily process for the first week, and then three days a week after that.

I stayed in the hospital for three weeks while the doctors tried to determine the cause of the disease and get me healthy enough to be on my own. After eliminating all other causes, they were left with excessive, long term hypertension, resulting in end-stage renal disease (ESRD). That confirmed the fact that dialysis was going to continue permanently.

The only other treatment for my condition is a transplant and that can take several years, if you qualify and if you survive. The survival rate for patients on dialysis drastically decreases after five years, exceeding the average wait time for a transplant in many hospitals across the country. This is not a pleasant reality, and in hindsight I clearly remember going through all the stages of grief throughout the next several months: denial, anger, bargaining, depression, and acceptance.

I had opened my first restaurant in 1991 and twenty years later I opened the second. Food, service, and business were all I had done in my adult life after finishing my graduate degree. Being in the kitchen was my comfort zone, and cooking was what I wanted to be doing.

The new place had been operating in my absence, but the original restaurant had not opened since the Friday before Labor Day weekend. I always closed during that three day weekend because the location was in the heart of the business district. On the Tuesday that I was scheduled to open again, I could not get out of bed for more than an hour at a time, and each time I did, I slept for the next two.

I told myself it must be exhaustion, but I am not sure I totally believed my own arguments. I knew that none of my friends were buying it and they had convinced me to make a doctor's appointment. I was scheduled to see him the Wednesday after the holiday.

The week before I went to see my doctor, I was feeling horrible. I had opened my second restaurant just about nine months before, but had been planning and building out the space for at least a year. The older business served breakfast and lunch and the new place served lunch, dinner, and late night. There were only a few hours each day that I was not working in one way or another.

My friend drove me to the doctor's office on Wednesday because I did not trust myself behind the wheel of a car. He waited for me, but he did not have to wait long. The nurse took my blood pressure and looked alarmed. She excused herself immediately and returned with the doctor.

He looked at me for a few seconds, retook my blood pressure, and told me I needed to go the hospital, immediately. The last thing I remember was passing through the waiting room doors in the emergency department into the treatment area. Then I awoke in the ICU.

I spent the next five and a half years on dialysis, and received a transplant in 2017. My kidney function is now good enough to keep me off dialysis. My restaurants closed down early on, and I was confronted with an entire new life.

This new life removed plenty of things from the old, but not my desire to cook and be in the kitchen. Although it took several months to navigate all the losses and understand the parameters of this new life, I was eventually able to find the energy to start cooking again, albeit for myself, and not for others.

There was a great deal to learn about the renal diet, restrictions of chronic kidney disease, and end-stage renal disease. The first of which is that they are not the same programs. Acceptable foods and amounts do not always apply for all stagesF of the disease. This also changes from patient to patient, even in the same grouping.

My nutritionists over those years all told me that CKD is the most difficult program to manage because of these variations, both for the patient and the nutritionist. The most common answer to any question from patient to nutritionist is " Well, it depends," and the more you learn about the disease and diet, it is also the most accurate.

I spent the next several years learning all I could from nutritionists, doctors, and even patients. They all provided me with valuable insight. What I did not find was information that satisfied me as a chef, foodie, or restaurateur, so I kept looking and asking questions. Doctors repeatedly told me they do not study nutrition in their coursework. The nutritionists, at best, were home cooks, and many even told me they do not cook at all. The food professionals, including myself at the time, did not have experience with the dietary needs of patients, and none had ever "sat in the chair." That chair is the dialysis chair, something you do not forget if you have spent any time sitting there.

In my opinion there was a great gaping hole in the information available for patients. This is the gap I have set out to fill. I listened to the professionals but wanted more exact answers. This lead me to deconstruct recipes that I had served to people over the years to determine the root of any problems, and if they could be altered to meet the needs of the CKD patient. I tested the recipes, detailed the nutritional components, and determined if any of the subsets of patients could consume the dish safely.

The first part of this book details the building blocks I needed to learn before I could continue this process. It provides information about the disease, it's progression, treatment, and the process of each level from a patient's perspective. It is everything I wish I knew when I was first diagnosed, and was not readily available. I hope it will help you understand the disease and the impact on the individual, their family, friends, and support systems.

The next section uses a cookbook format to provide recipes with detailed instructions. It includes notes about food, health, and cooking for every level of experience. After each recipe there are charts that identify the nutritional components for each ingredient, the recipe as a whole, and for each portion.

It is important to remember that not every recipe is suitable for every patient. It is exactly why this book was written. Every patient has different needs depending on their level of kidney function, and how they are processing certain nutrients and minerals. The charts may seem like an overload of information for a patient trying to manage this disease, but those charts can assist the care team with exact answers that are best suited for each patient.

The last section contains in-depth details about certain food groups, ingredients, and other important nutritional data. The information provides assistance with choosing better options for core ingredients used in this book's recipes or for your own favorites at home.

For those of us with chronic health issues, it is often difficult to explain to others the daily struggles we go through. There may be no outward physical appearance of illness on some days, or during some social outings, and we can appear to be healthy. The great effort it takes every day to manage routine activities, both physical and mental, go unnoticed. The only days people would see me were those when I had enough motivation and energy to get out of the house, and enough focus and mental acuity to engage in conversation. It generally did not last very long and I would leave as quickly as I felt all those factors begin to deplete. This can lead others to question, or even doubt, the difficulties we experience. Doctors and other professionals can be sympathetic and even empathize, but the people we see most often may not have an understanding of the full effects of the disease.

I have written this book because I have been on your path, I have sat in the chair with you, and I have had my life upended as you have. I have experienced the doubt of friends and family about the severity of my illness, and I have struggled on a daily basis to accomplish the simplest of goals. When I was diagnosed I did not know enough about this disease, nor did the people around me. This is my effort to assist those of you with chronic kidney disease and to provide a better understanding of it to those who provide care and assistance.

I hope the information compiled here, gathered from experts in different fields, can provide a better understanding of chronic kidney disease, and how the individuals, family members, and friends can assist in care.

Kidneys and CKD

What they do and how they work?

According to the National Kidney Foundation's website, our kidneys are bean shaped organs about the size of a fist that filter waste from the blood and remove it through our urine. They are "sophisticated trash collectors." In addition, they help produce red blood cells, maintain a balance of nutrients and some electrolytes, and keep our bones healthy in the process. Pretty important stuff.

The levels of phosphorus, potassium, vitamin C, and other minerals or nutrients, are monitored by the body and the kidneys help remove the excess. Sodium is an electrolyte and the kidneys assist in managing the amounts in our body by removing it through urine. Consuming too much salt can be very harmful, and if our intake is greater than the kidneys ability to remove it, health problems can ensue. The high salt western diet is a culprit for many of these issues.

Most of us have two working kidneys, but not all of us. It is possible to have both of your native kidneys and at some point have one stop working. You may not be aware of this fact until a medical scan, or some other health issue, reveals a size difference between the two. You can lead a normal life with only one working kidney, and have little or no side effects. This reality makes living organ donation for kidneys possible, with both donor and recipient living a healthy and normal life.

CKD: What is it?

Chronic kidney disease (CKD) is a progressive failure of the kidneys to filter toxins from the blood. The filtering process becomes less effective over time and the toxins linger in the blood much longer, causing health problems. There are a variety of causes of CKD including diabetes (Type I and II), high blood pressure, several scientifically named diseases (glomerulonephritis, interstitial nephritis, polycystic kidney disease, vesicoureteral reflux), and others.

Without getting into the medical terminology and description, partially because I am not qualified, here is a small description. The kidney is made up of a bunch of nephrons - the functional unit of the kidney. The nephron is made of two parts - the glomerulus and the tubule.

CKD occurs when the glomeruli (that's the plural) are not filtering the waste properly and the blood gets cleaned at lower levels. Your doctor may have told you about your GFR or Glomerular Filtration Rate, which indicates how well the kidneys are functioning. The number is in milliliters per minute (ml/min). This is referencing how much blood is filtering through the glomeruli per minute. The numbers look like they could be percentages, but they are not.

According to the DaVita website, those numbers have been broken down into stages:

Stage 1: GFR > 90ml/min	Normal kidney function
Stage 2: GFR = 60-89	Mild CKD
Stage 3: GFR = 30-59	Moderate CKD
Stage 4: GFR = 15-30	Severe CKD
Stage 5: GFR < 15	End-Stage Renal Disease

How does CKD effect the body?

There are several signs and symptoms of decreased kidney function, however, many of those can be associated with plenty of other health issues and are considered non-specific. The Mayo Clinic website suggests the following: nausea, vomiting, weakness, sleep issues. Some of the more distinct symptoms include changes in the amount you urinate, muscle twitches and cramps, swelling of the feet and ankles, persistent itching, difficulty controlling high blood pressure, and fluid build-up, which can cause chest pains and difficulty breathing. Regular visits to the doctor along with blood testing and blood pressure monitoring can assist in diagnosing CKD early.

What treatments are available?

In the early stages of chronic kidney disease the treatments can vary depending on the cause. The Mayo Clinic website suggests that treatment usually consists of controlling the signs and symptoms, reducing the complications that arise, and slowing the progression of the disease. In addition to the cause, symptoms are addressed using medications and often an altered diet to assist in slowing the progression of the disease.

If your kidneys reach ESRD, there are only two treatments available, dialysis and transplant. Medications are still used to treat the symptoms of your renal failure and the effects of dialysis, but sadly there is no cure.

Throughout most stages of CKD, patients' participation in their treatment is focused on compliance with appointments, medications, and following through with any physical and dietary recommendations. By the time the disease reaches end-stage, a great many things have been taken away from the patient. Dialysis, be it hemodialysis or peritoneal dialysis, can restrict the patient's freedom, and hence create a loss of control in their lives. Physical restrictions can limit normal daily activity, and the emotional drain can reduce the motivation to engage in some social activities, creating a feeling of isolation. All of these can accumulate and cause depression.

Having personally experienced most of the CKD issues mentioned so far, my focus here is to address one of the factors that we as patients do have some control over. Using my years of experience in the restaurant industry and my understanding of ingredients as a chef, I embarked on a journey to find ways to enjoy food while not adding to my current health problems. If I could find ways to assist the overall health of patients (myself included) and their quality of life, I thought the information should be shared.

What was my process for assessing and fixing the issues?

The first step was to learn about the problematic issues with food, not just for my own health, but for those with all levels of CKD. It is much more difficult to find the trouble makers in our food if we do not know what we are looking for. After learning all the dangerous dietary elements in my new life, I began by evaluating my current diet. Where is the potassium, sodium, phosphorus, protein, and other issues that my kidneys are not able to control? By documenting the food I ate over a period of time in a food diary, I was able to determine what I was actually eating versus what I thought I was eating. Once I established this, I deconstructed each recipe to find ways to make the end product better for my condition, and still appealing to my palette.

Writing down the recipes and amounts, something I almost never do at home, gave me the information I needed to evaluate the nutritional content of each meal. Next I would research each ingredient to gain a better understanding of the vitamins, minerals, and other nutrients per amount in the recipe.

Tallying up the results and determining my portion size gave me the starting point for each recipe. The next step was applying the scientific properties of certain ingredients as they affect the body; how they interact with each other throughout the cooking process, and any specific aspects related to CKD.

Finally I made adjustments I thought would work, tested out the recipes, and either included it, kept working on it, or just scraped the idea if it did not seem feasible. When I wanted to create a better program for others based on my research and experience, I came upon some interesting and helpful ideas.

Healthier eating, for any purpose, is a form of behavior modification. That is, changing someone's actions on a regular basis. This seems like an easy task for CKD patients given that the changes can provide a longer and better quality of life. However, this is not always the case. Just ask your friendly neighborhood psychologist or social worker. There are many factors involved and each person faces their own hurdles, whether social, economic, genetic, or learned. It is not as easy as you might think.

Using some adjusted techniques of behavior modification we can start with the following assessments. It is important to create a baseline of food for each individual. This is the food diary I described above. What are we eating and how often? In addition, it is not just the "what" part of eating, but where and how it is being prepared. There is big difference between a fast food restaurant burger or chicken sandwich, and one you make at home from fresh ingredients.

Part of that baseline includes frequency as well. How many different meals do you eat and how often? This is where things get interesting. There have been a few surveys that give us some feedback, but if you ask yourself, friends, and coworkers, I think the responses will support the results. In a survey in the United Kingdom reported by a food company wanting to find out how many different meals people ate and over what period of time, they found that 65% of people ate the same seven meals every week. The also found that of those, nearly half ate the same meals on the same day (Spillet, 2014).

Let's think about that for a moment. Almost 2/3 of the people surveyed have an eating cycle of seven days. If you assume they eat three meals a day, that's twenty-one meals each week. Now let's add in the idea that most people do not eat seven different meals for breakfast and seven different meals for lunch. The number of different meals is now dropping fast.

Not only did I ask friends about this breakdown, I vividly recall all my years owning restaurants and my interactions with customers. I knew most people by what they ate, and not always by their name. This is a common practice in the food service industry. Service staff and cooks usually know what their regulars are going to eat when they arrive. "Well done steak is here again at table 6," or "Everything on the side guy brought a new date tonight," are the type of remarks you might hear from the staff.

Using my experience along with the research stated above, it is my estimate that there are approximately four different lunch items and three different breakfast items, that each of us eat during the week. That would bring the total down to roughly fourteen different meals per week, give or take a few. The easiest way to support this is to ask yourself how many different meals you eat each week. This includes eating out and preparing food at home. You will find that this is an easily countable number. If you project out to ten days or even two weeks, I am willing to bet we can include nearly everyone in these food cycle parameters. The conclusion I am drawing here is that each of us has a very small number of meals in our cycle of eating. This implies that small changes in a person's diet can have significant impact on their nutrition. Once you have established this baseline, you can work to identify the problem areas and the possible changes that can be made.

It all sounded good on an individual level, but how could this be applied to the larger set of people with all the varieties in meals covering economic, cultural, and specific tastes? In order to apply this idea on a macro level, we find support in research conducted by General Electric. They needed to program cooking time and temperature for foods in a new oven using light and microwaves. The research they conducted found that 90% of all meals cooked could be programmed using only 80 recipes

(Wolke, 2010, p. 307). That is it, just 80. Combining these tw ideas tells us that not only do most people have a small rotati of meals on a weekly basis, we choose those meals from a sma set of recipes that cover 90% of the food everyone eats.

When I thought about implementing these ideas into the daily lives of CKD patients, I reverted back to my own patient perspective. How could someone convince me to not only try a new meal, but also to implement it into my routine?

I used the idea of ranking my meals in order of preference. If you allow the patient (in this case me) to rank the attractiveness of each meal, there might be a higher chance of participation. For instance, I would be less likely to remove my favorite regular meal, no matter how bad it may be, unless of course there was a noticeable and immediate detriment to my health. Even if I do, the best I could hope for is less frequent consumption. I found that this is much more significant in the CKD patient than the behavior modification attempts for other dieters.

The reason is the impact of the disease. In the initial stages you may not be feeling any large changes in your health or in your daily activities. And because it often progresses slowly the patient may not notice and be reluctant to make any changes, even with the knowledge that a dramatic outcome could be just around the corner.

If that dramatic outcome does happen and you end up with end-stage renal disease, spending time on dialysis, the effects of the treatment start to take its toll. So much gets taken away from the individual with little hope of correction or change, they may be less likely to add to those changes by willingly giving up their favorite foods or eating patterns.

Others can make drastic changes in their eating habits without looking back, but for many people, the transition is an ongoing battle. This is just another tool in attempting to aid each patient. They might even want to change out some of those meals, but are not aware of suitable replacements. There is a much better chance at success in altering behavior if the activity attempting to be changed is less desirable. So attack those meals on the bottom half of the ranking, for patients who cannot make those immediate and drastic changes.

What is included in the contents of this book?

Recipes

In this book you will find recipes that I have tested, and reconfigured when necessary, to meet the needs of patients with chronic kidney disease. Unlike many other sources of recipes, these have the benefit of being designed and presented by someone with over twenty years experience as a chef. As a patient, I used my time to learn and discover the nuances and differences in what are generally considered healthy eating habits and what is appropriate for CKD patients.

Having the opportunity to test many of these recipes on myself while on dialysis gave me a better understanding of the impact of diet on my CKD. Although I am just one data point, I was able to see the medical results through blood work, and gain a practical understanding of trying to live and stay healthy on dialysis. The physical, emotional, and social pains that accompany the treatment and the disease, aided my choices in food and process.

The effects of CKD at any stage are difficult to express. Chronic illnesses often have no visible symptoms and patients' issues can be overlooked even when verbalized. The care community - doctors, nurses, dietitians - may be able to sympathize, but a true understanding of the effects is much harder to absorb.

Several of the recipes have separate homemade ingredients included in the listings, such as aioli (mayonnaise, sort of), chili paste, and hot pepper sauce. Making the ingredients for a recipe may seem like a lot of work, but there are several uses for each of these specialty core ingredients throughout the book.

If you choose to purchase store bought items to replace these homemade ingredients, take note that the nutritional information will be altered in the final product. In each staple product, I point out the difference in homemade vs store-bought so that you are aware of the changes.

Recipe Table

After each recipe there is a chart of ingredients with their associated levels of nutrients, vitamins, and minerals that are of concern for CKD patients. In addition to the amounts, there are percentages of the total Reference Daily Intake (RDI) for each ingredient as provided by the U.S. Food and Drug Administration (USDA). Other countries have their own limits but they are often very close to those presented by the USDA. The RDI may not be the same for each patient, and if directed by your care team, you should make the appropriate adjustments. Having a general idea of what is recommended helps to understand the numbers. It took me quite some time to remember if 200 mg of phosphorus was a large amount or small. This is why I provide the percentages for each ingredient, the recipe as a whole, and each portion.

You may notice that some of the ingredients show very high levels of certain nutrients, vitamins, and minerals, however, the initial listing is for the entire dish. The end of each recipe lists the number of servings, portion size, and the appropriate amount for each serving. These final numbers are very important as each individual with CKD will have different concerns for their diet and health. It is important to talk to your care team, and especially the dietitian to better understand which items you may need to limit, and which you may need to increase.

Amino Acid Table

Proteins are comprised of amino acids. There are 20 amino acids that make up the proteins in your body. Nine of these amino acids need to come from outside the body, either through diet or supplements. These are considered the essential amino acids.

The second chart for each recipe contains a list of ingredients that contain protein, and the amounts of the nine essential amino acids they contain. It is not enough to get some of these amino acids, you must intake a specific amount of each per gram of protein in the food item. When you achieve this through one food item or in a recipe, the food item is considered a complete protein.

Nearly all meats, cheeses, and other dairy products will contain enough of all the essential amino acids per gram of protein to be considered a complete protein. The body needs proteins to build and repair tissue, produce hormones, enzymes, and other chemicals. If you are not getting all the essential amino acids in the right amount you can suffer from malnutrition and/or become severely ill.

CKD patients have different needs at different stages of their kidney failure and for the specifics of their body. Generally in the early stages (1-4), doctors will recommend decreasing overall protein consumption because processing proteins can be very taxing on the kidneys and can accelerate the overall damage.

If you are on dialysis, the process of removing the toxins and fluids from the body also removes some of the protein. This can cause a protein deficiency and as mentioned above, can be dangerous. At this point the care team will be watching your levels to ensure you are getting enough protein.

Protein intake and the essential amino acid content for recipes and ingredients can assist patients and care givers with applying the appropriate levels at the appropriate stages of CKD.

Other Charts

There is another chart in the index section that provides information about specific foods along with the protein and phosphorus levels. These numbers are a factor at all stages of CKD, but can take on a more important role if you reach ESRD and are placed on dialysis. Nearly all forms of protein intake come with high amounts of phosphorus. The body needs phosphorus mainly for the formation of bones and teeth, but also to make protein for tissue repair. The problem occurs when the body cannot control the proper amounts of phosphorus because the kidneys are not filtering out the excess.

This can lead to mineral deposits in your organs, muscles, and other parts of the body. It can also interfere with the processing of other vitamins and minerals. So we have a dilemma. How do you get enough protein without all the excess phosphorus? Many patients end up taking "binders" with their food. These pills will grab on to the phosphorus and not allow them to be absorbed into the body. They are helpful but the side effects can be difficult to manage. Gas, bloating, discomfort, diarrhea, and other problems often occur.

The chart in the index helps patients manage the phosphorus and protein levels associated with many different food items. It provides what is called the "Phosphorus to Protein" ratio for common protein based foods and their portion sizes. Ideally for ESRD you are looking for a low ratio, one where the amount of protein is high and the amount of phosphorus is low. It can also be helpful for patients in the early stages of CKD to identify the amounts of protein in certain foods, especially if you are on a protein limiting diet.

Inserts

In many of the recipes there are boxes of highlighted text with different colors. Each color represents a category for the type of comments. Green Boxes are food and cooking-related, blue is centered on health matters, and yellow is food science or a combination of the others.

The food and cooking notes provide some specifics related to the recipe where they are listed, and usually can be applied throughout the book. These tips include basic cooking skills and tips that can help the home cook better understand the impact of specific ingredients and processes on CKD. They also include advanced issues about food, its content, and how it reacts during cooking.

The health boxes contain information about how the body processes certain food items and how that relates to CKD. There are certain adjustments that can be made for better health through nutrition. Understanding the impact of certain foods on the body along with the makeup of the food we eat will advance the general understanding of all foods, and what you can do to stay healthier.

The food science boxes explain certain elements and properties of food, or are some combination of two areas that I thought should be separated.

Ingredients, food groups, and other information

Throughout my time on dialysis and post transplant, the most common problem I encountered was the overgeneralization of food items when categorized as positive or negative for patients with CKD. It is difficult enough to provide a general diet for CKD patients as each person has a different combination of ailments. Each patient also has a unique physiological response to both treatment and food balance.

Just as you cannot overgeneralize a diet for CKD patients, you should not do that with food groups either. Milk and cheese can vary significantly in their makeup based on production style or process. The contents of a frozen or processed food dish may appear to have the same ingredients as one you would cook at home, but the variances are often large and can be harmful. Throughout the book I point out the differences in food groups and recommend certain items (not brands) that can be beneficial to the patient and the reasons why.

This difference can apply to purchasing individual ingredients. An item as simple as ground beef can have a much different nutritional impact on the CKD patient, depending on the specifics, many of which are located on the package.

There are equipment and tools we use in the kitchen that make cooking easier and have an impact on the flavor and nutritional aspects of what we eat. Throughout the book there are descriptions and reasons for why I prefer certain items, and why I avoid others. Every cook has their own preferences in the kitchen, and I can always learn something, from professional chefs to home cooks, when it comes to tools and gadgets. This section also outlines some of the information I have gained over the years.

Vitamins/Minerals/Nutrients

As mentioned above, every CKD patient is going to have their own particular issues when it comes to controlling vitamins, minerals, and nutrients. Each recipe provides detailed information on the amounts and reference daily intake percentage using the recommended daily allowance guideline (how much you should consume). The data comes from the USDA databases, however, there can be slight variations in different products. Variables that can impact ingredients include growing regions, different crops, genetics, and processing for certain types of food. The information provided is a good guideline, but know that there may be some deviations.

Some of the more common items are potassium, phosphorus, protein, fats, and sodium. These items and others are listed, so be sure to check with your care team to identify which you should be monitoring.

There are also some general rules that may assist your food choices. Vitamins are either water soluble or fat soluble. This means that they are absorbed by the body in water or fat.

Vitamins A, D, E, and K are fat soluble; Vitamins B and C are water soluble.

If you need to increase your vitamin A levels and target foods that are rich in vitamin A, it will need fat to be absorbed.

One good example is a green salad. Green leaf lettuce is very high in vitamins A and K, however, if you use a fat-free dressing, your body will not be able to absorb the benefits of all those vitamins. Use a little olive oil and the absorption rate increases dramatically. The body does used stored fat to help absorb these items, but many patients suffer from malnutrition and have trouble maintaining a healthy weight.

The recipes also provide fat and saturated fat levels. Fats can be confusing for many of us as the different types have different effects on the body. We all need a certain amount of fat, and foods with unsaturated fats are recommended. It is necessary to point out that the argument over fat and health has been heating up for over twenty years. Is it good for you, neutral, or the cause of all that is evil in our health? I certainly have an opinion on the matter, but as with all CKD issues, the health aspects are highly dependent on each individual. What is no longer up for debate is eliminating trans fatty acids (trans-fats) from your diet. They are made by adding hydrogen to vegetable oils to make them solid. According to the American Heart Association's web site, trans fatty acids are no longer considered safe to eat.

High sodium levels in the body can also cause severe health problems even for people without kidney related issues. These include high blood pressure and increased fluid retention causing inflammation in the body. When too much sodium is ingested, the body acts to dilute the sodium levels by transferring water from the body's cells into the blood. This causes less water to be stored in the cells and triggers the body to drink more water to get back in balance. All the extra water puts a greater strain on the heart, kidneys, liver, and lungs. If you are not eliminating the extra fluid because of decreased kidney function, the effects are more problematic.

End-Stage Renal Disease (ESRD)

Stage 5, or ESRD, is a complicated subject and deserves further discussion. My five and a half years on dialysis gave me the opportunity to engage with many patients and caregivers, collecting a great deal of information about the process and systems. Although I am not a technical expert, I often assist others with the transition and learning curve as they find themselves going through this drastic change in their lives. Here are some of the things I wish I had known as I went though this unfortunate, but life saving change forced upon me.

There are two primary forms of dialysis, hemodialysis and peritoneal dialysis. Hemodialysis is conducted by removing excess fluid and toxins from your blood using an external filter, and returning it to your body. The process inserts two lines into a vein, one removes the blood and the second returns the blood. The first line feeds the blood into a machine which acts as much like your kidneys as technology allows. It then returns the cleaned blood through the second line.

There are two ways to access the blood in hemodialysis. The first is a catheter that is placed on one side of your chest, just below your collarbone. The catheter stays in your body between treatments and is almost always used when you start out on dialysis. There are two separate tubes that are covered and capped when not in use, and they need to be kept clean and monitored very closely. One end of each tube is internal and one is external to the body, making infection a risk. Since the location of the catheter is so close to the heart, any infection can be very dangerous. This is generally considered a short term solution because of this danger.

The second form of hemodialysis places two needles into a vein at the beginning of each treatment and removes them at the end. In order to do this a fistula or graft is placed either on the upper or lower arm, preferably of your non-dominant hand. The arteriovenous (AV) fistula is a plastic tube that is surgically inserted to connect a artery to a vein. This is done to increase blood flow and strengthen the vein where the needles are inserted. This is necessary as a normal vein could not withstand the constant access with the needles and could collapse. With an AV fistula, the needles are placed directly into the vein and over time the vein will increase in size and will protrude from the arm.

With an AV graft, a looped piece of plastic tube is surgically placed in the arm to connect the artery to the vein. The difference is that the needles are inserted into the looped plastic and not directly into a vein. The vein will not form the protrusions as with an AV fistula, but there are often problems with clotting and infection ("Vascular access to hemodialysis," n.d.).

Hemodialysis is generally performed as an in-center treatment, meaning you will go to a dialysis center three days a week for three to four hours of treatment. There are two different weekly schedules, Monday - Wednesday - Friday (MWF) or Tuesday - Thursday - Saturday (TTHS). Most clinics also have three shifts on each day. The starting times for patients in a shift are offset by 15- 30 minutes, to give the technician time to get each patient hooked up to their machine.

Home-hemodialysis (HHD) is also an option and requires both training and equipment placed in your home. Often you will need someone else to be at home with you or at least to be trained in the process in the event you are unable to cannulate yourself on any specific day. The needles are inserted by a trained professional or you can be taught to insert your own needles, into either an AV fistula or AV graft. One advantage is that you are not restricted to specific times and days for your treatment.

Peritoneal dialysis uses a solution inserted into the lining of the abdomen, or belly, to filter the blood of toxins and excess fluid. In order to do this a catheter in inserted on one side of the lower torso and remains there permanently. The fluid is inserted and then removed through the same catheter. This treatment is done by the patient and can be accomplished at home, work, or while traveling, as long the environment is clean.

The fluid remains in the belly for a few hours and then is emptied into the same bag that it started. The fluid is them discarded into a toilet or drain. While the fluid in inside your abdomen, you are free to go about your day as usual. This process needs to be done four to six times per day.

There is also a machine that allows you to do this at night, pumping the fluid in and out of your body as needed. You stay connected to the machine while this process is going on and it is done every night.

The fluid can make you feel bloated and because it is in you for most of the hours of the day, some people will need larger clothes ("Peritoneal Dialysis | NIDDK," 2018).

Both forms of dialysis will allow you to travel. With hemodialysis you will need to talk to the social worker at the clinic you attend or that oversees your home treatment. They will find available facilities in the area you are planning to visit, and make the necessary arrangements. For HHD patients, treatment during travel will depend on the type of equipment you use. The portable machines can be taken with you but if you are using one of the machines that is not portable you will have to get treatment at a center. The tricky part comes in the planning. Give the social worker at least a month to plan for your needs. If there is a facility owned by the company you currently use, that is where they will direct your treatment. If not, you will be directed to the nearest clinic that has a time slot available.

When you travel, the social worker will send a health history and treatment file to the facility you are visiting. This will be done prior to your travels. One of the problems many patients face is getting the needed information in time to make additional plans. You may not be informed of the location and especially the time until the day before, or often the day of, your travels. In addition you many end up switching the days of your treatment. This can be very discouraging and make it difficult to fully plan your time away. The reason is that many facilities will not know their available time slots and days until the day before you arrive. There are many variables that dictate chair availability in a center, so be patient.

Transplant

As mentioned earlier there are only two forms of treatment for ESRD. The first is dialysis and the second is a kidney transplant. Transplant is considered a treatment and never a cure, an important difference that should be understood. A successful transplant is considered a better option for treatment by most patients and caregivers. A positive result is that you will no longer have to be dialyzed. The goal is having the new kidney work well enough to clear out the toxins and fluids that your body was unable to do on it's own.

In addition, the transplanted kidney will be working twenty four hours a day. Whichever form of dialysis you were on was only cleaning the blood while your treatment was being conducted. In a three day a week program that equates to about 12 hours each week. That is a far cry from the 168 hours in a full week that working kidneys provide.

There are two categories of donors for a kidney transplant, living and deceased. A living donor is someone who volunteers to give you a kidney and live with one kidney themselves. This is most commonly a friend or relative. In recent years people who do not know the patient are offering to donate a kidney, these are called altruistic donors.

A deceased donor is someone who has just died and signed up to be an organ donor, or whose family agrees to have their organs donated. There is a limited time that organs can be removed and transplanted after the decision is made, meaning the recipient and the donated organ need to come together quickly. You will receive one kidney, and the other will go to a second patient.

The decision to seek a transplant as treatment has two main components. The first is the patient decision to pursue this option, and the second is the transplant facility's willingness to accept you as a patient. That may sound strange but it is necessary to go through both a physical and mental assessment to be accepted into a hospital's transplant program, regardless of the type of donor. The hospital will evaluate your overall health and make a decision. If they do not accept you into the program you should inquire what changes you could make for reconsideration.

Here are some unfortunate aspects for the process of getting on a transplant list. You must be healthy and cancer free for a certain period of time. Different facilities will have their own specific guidelines, but in general issues such as obesity, cancer, age, smoking, overall health, and a few other factors are going to play a role in the decision. One important factor is compliance. After a transplant, each patient will be taking some combination of immunosuppressants to make sure your body does not reject the new kidney. If a patient does not comply with their current doctor's directives by taking medications, or consistently getting their dialysis treatment, it is considered a warning sign that they may not comply with post transplant protocol, thus putting the new kidney at risk.

Once you get accepted at a transplant facility you will be put on the hospital's waiting list. There will many others on the list ahead of you and several factors determine how long you will have to wait. The list is based on blood type, time accumulated since being listed, need, and compatibility with potential donors (antigen match). The first variable is blood type. This is similar to the blood matching process when you need to get a transfusion. Your specific type will be capable of matching with certain other types, including your own. So the first item is to check for a compatible blood type. The list below explains how this works.

Your Blood Type	Who you can receive a kidney from
Type O	Only from another Type O
Type A	Type A and Type O
Type B	Type B, and Type O
Type AB	Type AB, Type A, Type B, Type O

Simplifying the information, you can receive a kidney from someone with your own blood type and from someone with Type O blood. Type AB can receive a kidney from another Type AB and from any other blood type, not just Type O. Because Type AB can receive a kidney from any other type, it is called the universal recipient.

The reverse of this chart is the donor chart of blood types.

Donor Blood Type	Who can receive a kidney from type
Type O	Type O, Type A, Type B, Type AB
Type A	Type A, Type AB
Type B	Type B, Type AB
Type AB	Type AB

This chart shows that you can give a kidney to someone with your own blood type plus anyone with Type AB blood. Type O donors can give a kidney to any blood type, and are called the universal donor.

This blood type matching has a big impact on how long you will wait on a list. The breakdown of individuals with different types of blood is as follows (Cicetti, 2012):

Type O	45%
Type A	40%
Type B	11%
Type AB	4%

Without going into too much math let's look at how this affects a few types. If you are Type O, there are around 45% of donors that will have the blood type you need. That means you can potentially receive a kidney from about 45% of individuals. That's the good news. The negative side is that all other blood types can accept Type O blood, so they are also eligible to receive a kidney from a Type O blood donor. If a Type O kidney becomes available, anyone with more time on the list can be considered for this kidney. Simply put, you can only receive a kidney from a similar blood type, but at least almost half of people have that type. But there is also a higher demand for that 45% since any blood type can accept Type O.

On the other hand if you are Type AB, you are capable of accepting a kidney from 100% of blood types, giving you a greater chance to match. Even though there are only around 4% of the population with Type AB, and there will be fewer people waiting with that blood type. They have a higher probability of getting a kidney faster if all other things are equal.

On the donor side, if you have Type O blood, you can give to anyone, and have a higher chance to match up with a potential recipient. However if you are Type AB, you can only donate to another Type AB, and this will be more difficult to find.

Confusing? Yes. But it's not an easy process to match a kidney, and this is just one of the variables. There are a few studies currently being conducted to assist with this type of matching, and as time goes on and the research continues, the numbers game may even out and become less of a factor and blood type may become less of a concern.

Time

If you have a living donor, the time on a list is not an issue. It will take about six months to complete all the tests and prepare for the transplant with a living donor, so be patient.

If you are waiting for a deceased donor kidney, time on the list is more of a factor. The more time you have on the list the less time it will take to be considered for matching. There is not much you can do to increase the amount of time, and "the squeaky wheel" theory is not an element of this process.

The United Network of Organ Services (UNOS) is a national organization that oversees the organ donation system in the US. They keep the data on people in need of transplants and on the matching criteria. This is one of the few organizations that does not appear to be affected by outside influences, including power, money, or fame.

You may often hear of a celebrity getting a kidney right after being diagnosed and wonder if they are circumventing the system, but it is always because they have a living donor willing to provide a kidney. The greater the celebrity, the more people might be willing to offer.

There are ways to help yourself with the waiting time for a deceased donor kidney. Since time is limited once a deceased donor kidney is available to be transplanted, the country is broken into sections called procurement regions, and each region is given first priority when an organ becomes available from their area. Even if there are several hospitals in the same region you may be limited to listing in only one from that specific region as you are drawing from the same available organs.

You can get listed in different regions if you meet the criteria for the different hospitals. This is something you should do immediately if you are looking at this option. Multiple listings will increase your chances of getting a kidney faster.

There are a few websites that can tell you how long the wait time is in each hospital based on your blood type. This information is extremely helpful if you have the ability to travel for your transplant. One website for this information is http://www.txmultilisting.com/wait.htm.

This website compares all the transplant facilities for wait times. You can choose a specific hospital, region, or time, and it provides you results right away. You can also just enter your blood type, and it shows the hospitals listed by wait times in months. The difference can be very large, ranging from 10 to over 300 months, depending on your blood type and the hospital.

There are also websites that rank kidney transplant hospitals. This should be a factor as well but understand that the range of quality is not enormous, as the government oversees these hospitals and they have strict guidelines to keep their accreditation and to continue to transplant kidneys. You can find this information on websites such as https://www.healio.com.

One question transplant centers will ask when you apply to be listed in an area outside your own region concerns support, financial, social, and emotional. Once you receive a transplant you spend a few days to a few weeks in the hospital, however, you will need to return quite often over the next 6 to 8 weeks. During this time you could be required to return to the hospital every day for a period of time, and then to slowly decrease the visits. You will not be allowed to drive yourself either.

This means you will need a place to stay, transportation in the area, and a support system that will assist with your care while you recuperate. There are lower cost facilities that can assist with housing for you and family members, but there will be costs associated with receiving a transplant outside your home area vs. one close to home.

Compatibility

Compatibility, as mentioned above, includes blood typing but it is not the only variable that is compared when determining if a donor kidney can be transplanted. The National Kidney Foundation's website gives a good overview of the tests that are done after blood typing to determine if you and a potential donor are a match for transplantation. These tests evaluate the bodies ability to accept the new kidney by looking for compatible genetic markers. Some issues that can affect compatibility are previous blood transfusions, pregnancy, and past transplants.

All of these tests help reduce the risk of rejection. This is when the body views the transplanted organ as a foreign object and the immune system attempts to kill the invader. The same thing occurs with viruses and bacteria. Keep in mind that your body could still reject the new kidney as the system is not 100% accurate. Hospitals can face difficulties from governing bodies if the rates dip below certain levels. You can find out the hospitals success rate by asking during your evaluation, if they have not already given you paperwork that has that information.

Do not get too bogged down with many of these details as the hospital will be doing all the testing and there is nothing you can do to alter the results. It is important to know that blood type is not the only factor and you may find that some living or deceased donors will not be a match for you.

Recipes

The recipe section is not organized in a typical cookbook fashion. You will find that others are organized into meal types like breakfast, lunch, dinner, dessert, and snacks. Some are grouped according to the content of the meal such as meat, soup, pasta, appetizers, or some other format.

My collection of recipes presents information about cooking and the important factors of kidney disease in an order that is meant to be educational and additive. Each recipe provides information on food and the connection to CKD, along with a delicious end product. The order is meant to provide a variety of food choices along the way, and to build your knowledge of cooking, food, and CKD.

Olio Grezzo Naturale
Giovannella
Olio Extra
Vergine di Oliva
Contenuto 1000 ml ℮
Non Filtrato ~ Unfiltered

Lemon Parmigiano Pasta

I wanted to start this section with one of my favorite dishes. It is simple to prepare and highlights how fresh ingredients can have a big impact on flavor. I was preparing this meal long before my diagnosis and realized it was one dish that I did not need to replace or adjust.

Ingredients:

- 1/2 pound long pasta (spaghetti, fettuccine, etc.)
- 1/2 cup onion, chopped
- 3 lemons, zest of 2, and juice of 2 1/2, save the other half for garnish
- 1/3 cup olive oil
- 1/4 cup fresh parsley
- 1/3 cup freshly grated Parmigiano Reggiano
- 2 teaspoons salt

Total Time: 30 minutes Yield: 18 oz. Portions: 4
Serving size: 4.5 oz.

Fill a large pot with a little more than 2 quarts of water and place it on high heat on the stove. A good rule here is 1 gallon of water to 1 pound of pasta.

Next, heat a skillet, large enough to hold all of the cooked pasta, on a medium setting. We are going to start by sweating the onions. That means slowly cook them in fat, until they are translucent. You can also sweat until soft, but the goal here is not to brown them.

The next two steps you can do at the same time if you are feeling comfortable and have all your ingredients ready to use. Lemons zested and juiced, onions chopped, and cheese grated.

When the pot is boiling, add the pasta and 2 teaspoons of salt to the water. Stir to make sure none of the pasta is sticking to the bottom of the pot. Pasta will take about 8 minutes to cook, depending on it's size and thickness. Check your label for recommendations.
To the pan: Once the pan is hot, add the olive oil and wait about 15 seconds until the oil warms. Add the onions, moving them

around often until they become translucent, 3 to 5 minutes. Now add lemon zest and juice of 2 1/2 lemons, reserving the other 1/2 lemon for a garnish. Let this mixture cook for about 3 minutes on a low to medium heat. Here we are infusing the oil further with the lemon flavor.

<u>Food Note</u>

Sweating brings out some of the water in the veggies, and with it some of the other flavors that reside inside. This flavor will transfer to the fat, and the chemical makeup of the veggie will change, thus changing the flavor. In onions, this will reduce the strong raw onion flavor and move the sugars in the veggie to the forefront, making the final product sweeter. Sautéing is done at a higher heat and browns the veggie.

<u>To the Pot:</u> Cook the pasta until almost done (al dente). Using a pasta prong or set of tongs, grab the pasta from the water, let it drain as much as possible back into the pot and then add directly to the oil mixture. This lets you reserve some of the starchy water in case you want to add a little pasta water to moisten up the dish. Next, add the parsley and mix thoroughly. Now turn off the heat and add the Parmigiano Reggiano and mix again.

Transfer the pasta to a serving bowl, and garnish with lemon slices and parsley.

<u>Food Note</u>

Let's talk about cooked pasta. Once the pasta is cooked (soft but a little firm) you have a few options. Draining your pasta will remove most of the water and some of the starch. The starch is what makes any sauce stick to the pasta. It can also make the pasta stick to itself if left un-sauced. Rinsing your pasta will remove the starch and your sauce will not stick to the pasta. Removing the pasta directly from the water and placing it in the sauce helps reserve the starch needed to adhere to the pasta.

Health Note

An article in the BBC news identifies the effect of pasta on blood glucose. These levels were tested from volunteers 2 hours after eating a dish of pasta. Each time the same individuals were tested. The first test was conducted after eating pasta directly after it was cooked and the second test was conducted after eating the cooked pasta, cold, after it had been in the refrigerator overnight. The results showed that there was a 7% decrease in blood glucose levels after eating the cold pasta. The third test was performed using the cold pasta, and reheating it before eating. The results showed that the blood glucose levels were 48% lower after eating cold pasta.
The science is is also suggested by Montignac ("factors that modify glycemic indexes I Official web site of the Montignac Method," n.d.), showing that pasta is a resistant starch and the reheating changes how the enzymes react in the body. After reheating, they acted more like fiber than carbohydrates, lowering the glycemic index level.

Lemon Parm Pasta

		Fat (g)	Sat. Fat (g)	Phos (mg)	Pot (mg)	Sodium (mg)	Calories	Carbs (g)	Vit A (iu)	Vit C (mg)	Vit K (Mcg)	Vit E (mg) AT	Vit B6 (mg)
	RDI	**65**	**22**	**700**	**3300**	**2300**	**2000**	**275**	**5000**	**60**	**120**	**20**	**60**
Ingredients	**Amt**												
Pasta	**1 lb**	1	0	182.4	86.4	337.6	1600	104	0	0	0	0	0
		1.54%	0.00%	26.06%	2.62%	14.68%	80.00%	37.82%	0.00%	0.00%	0.00%	0.00%	0.00%
Lemons 3 Zest and juice	**6 Tbsps**	0	0	9.4	174.9	1.5	35.1	9	28.2	198.6	1.6	0	0
		0.00%	0.00%	1.34%	5.30%	0.07%	1.76%	3.27%	0.56%	64.80%	1.33%	0.00%	0.00%
Parmigiano Reggiano	**1 oz (1/4 cup)**	7	5	172	13.2	428	121	1	112	0	0.5	1.2	0
		10.77%	22.73%	24.57%	0.40%	18.61%	6.05%	0.36%	2.24%	0.00%	0.42%	6.00%	0.00%
Olive oil	**1/3 cup**	72	10	0	1	1	442	0	0	0	30.1	7.2	0
		110.77'	45.45%	0.00%	0.03%	0.04%	22.10%	0.00%	0.00%	0.00%	25.08%	36.00%	0.00%
Fresh Parsley	**1/4 cup**	0	0	18.1	77	7.6	0	0	1179.5	18.6	230	0.1	0
		0.00%	0.00%	2.59%	2.33%	0.33%	0.00%	0.00%	23.59?	31.00%	191.67%	0.50%	0.00%
Onion Chopped	**1/2 cup**	0	0	23.2	117	3.2	4	0	1.6	5.9	0.3	0	0.1
		0.00%	0.00%	3.31%	3.55%	0.14%	0.20%	0.00%	0.03%	9.83%	0.25%	0.00%	0.17%
Salt	**1 tsp**	0	0	0	0	1500	0	0	0	0	0	0	0
		0.00%	0.00%	0.00%	0.00%	65.22%	0.00%	0.00%	0.00%	0.00%	0.00%	0.00%	0.00%
Total for Dish		80	15	405.1	469.5	2278.9	2202.1	114	1321.3	223.1	262.5	8.5	0.1
		123.08'	68.18%	57.87%	14.23?	99.08%	110.11%	41.45%	26.43?	371.83?	218.75%	42.50%	0.17%
# of Servings		8	8	8	8	8	8	8	8	8	8	8	8
Totals per serving		10.00	1.88	50.64	58.69	284.86	275.26	14.25	165.16	27.89	32.81	1.06	0.01
		15.38%	8.52%	7.23%	1.78%	12.39%	13.76%	5.18%	3.30%	46.48%	27.34%	5.31%	0.02%

Essential Amino Acid Chart for recipe Lemon Parm. Pasta

	Mg needed per gram of Protein	Lemons Juice & Zest	Recipe 6.1 g	Pasta no Egg	Recipe 56 g	Parm. Regg.	Recipe 10 g	Olive Oil	Recipe .4 g	Onion Chopped	Recipe .9 g	Total Grams	Total Amino Acid	Total Protein/g
Essential Amino Acids		Protein/g	Recipe	Protein/g	Recipe	Protein/g	Recipe	Protein/g	Recipe	Protein/g	Recipe			
Histidine	18	16.90	103.09	20.28	1135.68	38.70	387.00	27.47	10.99	12.44	11.20	73.4	1647.96	22.45
Isoleuciine	25	36.86	224.83	38.52	2157.12	53.00	530.00	61.05	24.42	11.79	10.61		2946.98	40.15
Methionine	25	11.48	70.00	15.52	869.12	26.80	268.00	25.11	10.04	1.68	1.52		1218.68	16.60
Leucine	55	44.29	270.14	68.00	3808.00	96.70	967.00	98.95	39.58	21.05	18.95		5103.67	69.53
Lysine	51	47.62	290.48	19.16	1072.96	92.58	925.84	80.00	32.00	32.84	29.56		2350.84	32.03
Phenylalanine	47	28.38	173.12	48.40	2710.40	53.80	538.00	49.00	19.60	22.22	20.00		3461.12	47.15
Threonine	27	30.76	187.65	26.44	1480.64	36.90	369.00	45.42	18.17	17.68	15.92		2071.37	28.22
Tryptophan	7	9.81	59.84	12.76	714.56	13.50	135.00	14.32	5.73	12.44	11.20		926.32	12.62
Valine	32	43.29	264.04	42.40	2374.40	68.70	687.00	67.89	27.16	18.67	16.80		3369.40	45.90
Amt. per recipe		3 each		1 lb		1 ouce		4 tbsp		1/2 cup				
Prot. per recipe		6.1		56		10		0.4		0.9				

Total recipe Protein (g)	# of servings	Protein / serving (g)	Not a Complete Protein
73.4	8	9.175	

Hot Pepper Sauce

Hot sauce is a great flavor enhancer for any food and if you have CKD, it can give a boost of flavor in some of the bland recipes you may have found in other sources. One major drawback to many of the commercially made hot sauces is the amount of added salt.

Most bottles give a portion size of 1 teaspoon and the nutrition information for that portion. For some of the spicy sauces this may be enough. But for many others this amount underestimates actual usage. This can be helpful to the manufactures for labeling as they are high in sodium. In my experience as a restauranteur, many dishes you consume will have quite a bit more that that 1 teaspoon. An easy example is the traditional Buffalo chicken wing sauce. Even 1/2 dozen chicken wings will use about 2 tablespoons of hot sauce and can put sodium intake well over the recommended daily amount of 1500 to 2300 mg.

Ingredients:

- 18 fresno, cayenne, or jalapeño peppers. (The color of the peppers determines the color of hot sauce).
- 2 additional hotter peppers, i.e. jalapeños
- 2 cups white wine vinegar or cider vinegar (Not that distilled stuff)
- Juice of one whole lime, about 2 tablespoons
- 3 garlic cloves
- 1 shredded carrot
- 1 teaspoon salt

Total Time: 45 minutes Yield: About 1 pint if you remove the seeds through a strainer. Up to a pint and half without straining
Portions: 16-24 Portion size: 2 tablespoons

Equipment: Blender, Immersion blender, or Food Processor

Start by cutting off the stems of all the peppers and discarding. Next cut each pepper into 4 to 6 pieces so that it is a bit quicker and easier to blend.

Cut the garlic cloves into a few smaller pieces and combine the vinegar, lime juice, garlic, and all the peppers (seeds included) into a pot. Bring to a boil and reduce heat to a simmer. Cook for 20 minutes.

Food Note

Hot peppers hold their heat in what is called the pith. This is the white lining that is found between the ribs on the inside of the pepper and on the top part holding the seeds. It is a common misstatement that the heat is in the seeds, but that is because some of the pith left on seeds causes the heat. If you want to reduce the heat, remove more of the pith before using the pepper.

Remove the pot from the stove and let cool, or place in the refrigerator for 15 minutes. Hot food in blenders or food processors can be very dangerous.

When the pepper mixture has cooled to room temperature, fill your blender or food processor to just under the half-way mark. You will need room for the mixture to combine and spin.

Mix very well, until all the peppers have combined with the vinegar and you don't see many chunks. The seeds will not grind up, but that's OK.

Remove pepper sauce, and repeat with the rest of the peppers. At this point you will have a choice to make. You can either leave the seeds in for a sauce with more texture, or use a strainer for a smooth sauce.

Straining the hot sauce gives it a smooth appearance as you remove large pieces and the seeds. Some like the appearance of the seeds and pieces along with the extra heat from any pith left over that comes when the sauce is not strained.

You can use the sauce as soon as you are finished, but a few days in the fridge will enhance the flavor. In addition, the hot sauce will last for months if kept in the fridge as it is a vinegar based product.

You will notice that I did not use the salt listed in the recipe. The nutrition charts below show the recipe with the salt. I prefer to add just a little sea salt to whatever I am putting the hot sauce on to enhance the flavor, if needed. By doing this I will know exactly how much sodium is going in my food. Cook's choice on adding to the recipe.

Health Note

Capsaicin! Just Wow. Capsaicin is the primary active ingredient in chili peppers and has been used for centuries in the treatment for many medical issues. In recent studies (Zheng, 2017) to prove the accuracy of these traditions, results have been given a thumbs up in many areas. These include pain, inflammation, weight loss, as an antioxidant, and as a treatment for neurodegenerative diseases. Like I said, wow. We all can benefit from this kind of help, but for CKD patients, the anti-inflammatory and pain relief properties are of particular interest ("Compound found in chili peppers," 2013)..

Homemade Hot Pepper Sauce

		Fat (g)	Sat. Fat (g)	Phos (mg)	Pot (mg)	Sodium (mg)	Calories	Carbs (g)	Vit A (iu)	Vit C (mg)	Vit K (Mcg)	Vit E (mg) AT	Vit B6 (mg)
	RDA	**65**	**22**	**700**	**3300**	**2300**	**2000**	**275**	**5000**	**60**	**120**	**20**	**2**
Ingredients	**Amt**												
Peppers. Fresno & Jalapeños	**20**	0	0	294	2254	56	280	54	6671	1008	98	4.9	3.5
		0.00%	0.00%	42.00%	68.30%	2.43%	14.00%	19.64%	133.42	1680.00	81.67%	24.50%	175.00
Apple Cider Vinegar	**2 Cups**	0	0	38.2	348	24	100	4	0	0	0	0	0
		0.00%	0.00%	5.46%	10.55%	1.04%	5.00%	1.45%	0.00%	64.80%	0.00%	0.00%	0.00%
Lime Juice/ Zest	**1 oz**	0.00	0.00	8.00	32.80	1.20	14.00	4.00	14.00	8.40	0.20	0.10	0.00
(2 Tbsp)		0.00%	0.00%	1.14%	0.99%	0.05%	0.70%	1.45%	0.28%	14.00%	0.17%	0.50%	0.00%
Carrots - Peeled	**1 Med**	0.00	0.00	38.50	352.00	75.90	52.00	8.00	21363.	7.60	16.90	0.80	0.20
(61 g)		0.00%	0.00%	5.50%	10.67%	3.30%	2.60%	2.91%	427.26	12.67%	14.08%	4.00%	10.00%
Garlic clove	**1 med**	0.00	0.00	4.60	12.03	0.50	4.33	1.00	0.27	0.93	0.07	0.00	0.03
		0.00%	0.00%	0.66%	0.36%	0.02%	0.22%	0.36%	0.01%	1.56%	0.06%	0.00%	1.67%
Salt	**1 Tsp**	0.00	0.00			2300.00							
		0.00%	0.00%	0.00%	0.00%	100.00%	0.00%	0.00%	0.00%	0.00%	0.00%	0.00%	0.00%
Total for Dish		0.00	0.00	383.30	2998.83	2457.60	450.33	71.00	28048.	1024.93	115.17	5.80	3.73
		0.00%	0.00%	54.76%	90.87%	106.85%	22.52%	25.82%	560.97	1708.22	95.97%	29.00%	186.67
# of Servings		16	16	16	16	16	16	16	16	16	16	16	16
Totals per serving		0.00	0.00	23.96	187.43	153.60	28.15	4.44	1753.0	64.06	7.20	0.36	0.23
		0.00%	0.00%	3.42%	5.68%	6.68%	1.41%	1.61%	35.06	106.76	6.00%	1.81%	11.67%

Essential Amino Acid Chart for recipe Red Hot Pepper Sauce

Essential Amino Acids	Mg/ gram protein needed	Peppers Hot	Recipe 6.1 g	Apple Cider Vinegar	Recipe 56 g	Lime - Juice & Zest	Recipe 10 g	Fresh Garlic	Recipe 2.7 g	Carrots Peeled into strips	Recipe .8 g	Total Gra ms	Total Amino Acid	Total Protein /g
		Protein/g	Recipe	Protein/g	Recipe	Protein/g	Recipe	Protein/g	Recipe	Protein/g				
Histidine	18	21.58	131.63	0.00	0.00	38.70	3.87	17.00	10.20	21.67	17.33	7.6	163.03	21.45
Isoleuciine	25	34.21	208.68	0.00	0.00	53.00	5.30	32.50	19.50	56.67	45.33		278.82	36.69
Methionine	25	12.63	77.05	0.00	0.00	26.80	2.68	11.50	6.90	10.00	8.00		94.63	12.45
Leucine	55	55.26	337.11	0.00	0.00	96.70	9.67	47.00	28.20	58.33	46.67		421.64	55.48
Lysine	51	46.84	285.74	0.00	0.00	92.58	9.26	42.00	25.20	55.00	44.00		364.20	47.92
Phenylalanine	47	32.63	199.05	0.00	0.00	53.80	5.38	27.50	16.50	43.33	34.67		255.60	33.63
Threonine	27	38.95	237.58	0.00	0.00	36.90	3.69	23.50	14.10	51.67	41.33		296.70	39.04
Tryptophan	7	13.68	83.47	0.00	0.00	13.50	1.35	10.00	6.00	15.00	12.00		102.82	13.53
Valine	32	44.21	269.68	0.00	0.00	68.70	6.87	43.50	26.10	60.00	48.00		350.65	46.14
Recipe Amount		20 each		2 cups		1 each		3 clove		1 cup				
Protein per recipe (g)		6.1		0		0.1		0.6		0.8				

Total recipe Protein (g)	# of servings	Protein / serving (g)	Not a Complete Protein
7.6	16	0.475	

Desserts

Food Note

With an already confusing and restrictive diet, desserts can be especially difficult for the CKD patient. Calories, sugar content, fats, and phosphorus are all major components of many desserts. These are also items that may need to be controlled by someone with CKD.

Fats, both vegetable oils and animal fats (think canola oil and butter) are necessary when making most desserts. Both provide moisture, lubrication when blending ingredients, flavor, and mouth feel. High levels of saturated fat can have detrimental impacts on certain at-risk patients.

The creation of partially hydrogenated oils - turning oils from liquid to a solid at room temperature - had a big impact on most commercial bakeries. These are the trans-fatty acids that we all hear about. They were cheaper, more stable, and used in most bakery products and desserts. It the mid 1990's there was a movement to recognize the health risks of trans-fatty acids. It was not until 2013 that they were recognized by the FDA as not safe, and not until June of 2018 that their use was outlawed in commercial baking. There are some exceptions so it is important to check your labels and avoid the usage of products that include them.

The nutritional issues for CKD patients are a balancing act, especially in desserts. Talk to your dietitian to make sure you are aware of the dangers and risk factors you may face when eating these products.

Homemade Apple Sauce

As with several recipes in this book, you might ask "Why bother?" With all the options that are produced and available it would be much easier to buy it at the store. Well, there are several good reasons, but the best one I can think of is the taste. Once you tasted the homemade stuff, no further argument should be necessary.

For the CKD patient, increasing the taste of food without adding things that are bad for us is highly beneficial and recommended. Whether it is a warm bowl of apple sauce on a cool night, right from the fridge on hot days, or as a replacement for some fats in baked goods, the homemade variety is a much better choice.

The next argument is standard for any food. The less processing a food item goes through the better the end result. Fresh is best, frozen at the peek of flavor is good, and the least amount of machine processing is the better option if the others are not available.

This is a very simple and quick recipe with exceptional results.

Ingredients:

- 4 apples, cored and sliced into pieces (skin on is better, but peeled will work)
- 1 1/2 cups apple cider (or 1 cup each of brown sugar and water)
- 1/4 cup fresh lemon juice
- 2 cinnamon sticks broken into 1/3's (or 1 tablespoon ground)
- 1 tablespoon cognac or brandy (optional)

Time: 30 minutes   Yield : 4 cups   Portion size 1/2 cup

Portions: 8

Preparation is fairly simple as the recipe suggested, however, there are some options.

Once you have prepped all the ingredients place them in a pot and cook on medium heat until the sauce starts to thicken. Lower heat to medium low until the apples are soft and tender.

Use a masher to break up the the apple pieces. If you left the skins on the apples and want them removed, you can run the apple sauce through a food mill or large strainer. The other option is to peel the apples before you start cooking.
Serve warm or refrigerate for up to a week to 10 days, if it lasts that long.

Let's discuss the skin of the apple. It contains most of the vitamins and about half of the fiber. Leaving the skin on increases the healthy benefits of apples. We have all heard the (possibly) Welsh phrase "An apple a day, no doctor to pay" or it's other derivatives.

If you can enjoy the apple sauce with the skin, it is a very good boost for your health, and it's quite versatile.

Health Note

Pectin is a soluble fiber that is found in ripe fruits. It is used as a thickening agent in cooking and baking. The health benefit of soluble fiber is in its ability to raise the high density lipoproteins (HDL's) in the body. This is the good cholesterol. Pectin also reduces the low density lipoproteins (LDL's), the bad cholesterol. Cholesterol is carried in the blood by attaching to lipoproteins. The HDL's act in a way to eliminate the excess cholesterol in the body by attaching and then removing excess through waste. LDL attaches to cholesterol, but hangs around, even if the cholesterol load is excessive in the body.

The alcohol in the brandy or cognac will burn off, and the concentration of the flavors will add a little zing to the apple sauce. It is optional and will still taste very good without it.

The other option is to add a few tablespoons of butter, as many recipes suggest. In this case I leave out the fat mainly because the amount of fat soluble vitamins is low, and the addition gives the end product a nice shine, but not much flavor.

Apple Sauce

		Fat (g)	Sat. Fat	Phos (mg)	Pot (mg)	Sodium (mg)	Calories	Carb (g)	Vit A (iu)	Vit C (mg)	Vit K (Mcg)	Vit E (mg) AT	Vit B6 (mg)
	RDA	65	20	700	3300	2300	2000	110	5000	60	120	20	2
Ingredients	Amt												
Apples (750 g)	6 Cups	0	0	82.8	804	7.8	390	102	405	40.2	16.8	1.2	0.6
4 Medium		0.00%	0.00%	11.83%	24.36%	0.34%	19.50%	92.73%	8.10%	67.00%	14.00%	6.00%	30.00%
Apple Cider	1 1/2 cup	0	0	26.1	376	15	221	42	3.75	3.30	0	0	0
(248 g)		0.00%	0.00%	3.73%	11.39%	0.65%	54.00%		0.08%	64.80%	0.00%	0.00%	0.00%
Lemon Juice - fresh	1/4 cup	0.00	0.00	27.60	72.20	3.00	26.00	6.00	1.60	5.60	0.40	0.00	0.20
(2 Tbsp)		0.00%	0.00%	3.99%	2.19%	0.60%	1.30%	5.45%	0.03%	9.33%	0.33%	0.00%	10.00%
Cinnamon Stick	2 each	0.00	0.00	10.00	66.80	1.60	38.00	16.00	45.80	0.60	4.80	0.40	0.00
1 Tbsp Ground		0.00%	0.00%	1.43%	2.02%	0.07%	1.90%	14.55%	0.92%	1.00%	4.00%	2.00%	0.00%
Cognac`	1 Tbsp	0	0	0	0	0	0	0	0	0	0	0	0
		0.00%	0.00%	0.00%	0.00%	0.00%	0.00%	0.00%	0.00%	0.00%	0.00%	0.00%	0.00%
Total for Dish		0.00	0.00	146.50	1319.00	27.40	675.00	166.00	456.15	49.70	22.00	1.60	0.80
		0.00%	0.00%	20.93%	39.97%	1.19%	33.75%	150.91	9.12%	82.83%	18.33%	8.00%	40.00%
# of Servings		8	8	8	8	8	8	8	8	8	8	8	8
Totals per serving		0.00	0.00	18.31	164.88	3.43	84.38	20.75	57.02	6.21	2.75	0.20	0.10
		0.00%	0.00%	2.62%	5.00%	0.15%	4.22%	18.86%	1.14%	10.35%	2.29%	1.00%	5.00%

Essential Amino Acid Chart for recipe Applesauce

	Mg/ gram protein needed	Apples with skin	Recipe 1.2 g	Apple Cider	Recipe 2.7 g	Lemon Juice	Recipe .2 g	Total Grams	Total Amino Acid	Total Protein/g
Essential Amino Acids		Protein/g	Recipe	Protein/g	Recipe	Protein/g	Recipe			
Histidine	18	21.00	25.20	21.00	56.70	38.70	7.74	4.1	89.64	21.86
Isoleuciine	25	25.00	30.00	25.00	67.50	53.00	10.60		108.10	26.37
Methionine	25	4.33	5.20	4.33	11.70	26.80	5.36		22.26	5.43
Leucine	55	54.33	65.20	54.33	146.70	96.70	19.34		231.24	56.40
Lysine	51	50.00	60.00	50.00	135.00	92.58	18.52		213.52	52.08
Phenylalanine	47	25.00	30.00	25.00	67.50	53.80	10.76		108.26	26.40
Threonine	27	25.00	30.00	25.00	67.50	36.90	7.38		104.88	25.58
Tryptophan	7	4.33	5.20	4.33	11.70	13.50	2.70		19.60	4.78
Valine	32	50.00	60.00	50.00	135.00	68.70	13.74		208.74	50.91
Total Amount		6 Cups	750 g	1 1/2 cups		4 Tbsp				
Protein per recipe (g)		1.2		2.7		0.2				

Total recipe Protein (g)	# of servings	Protein / serving (g)	Not a Complete Protein
4.1	8	0.51	

Carrot Cake

Carrot cake is delicious. I did not always think so, or at least I could not get over the name, so I rarely tried it. I really enjoy it now, even though carrots are not always on the list of acceptable vegetables for CKD patients. High potassium levels in carrots make them risky for CKD patients. However, the recipe below shows that the per serving potassium level should not pose a risk for most patients. There are additional changes from a standard carrot cake recipe that help out the other problem issues as well.

Health Note

This shows the effect of ingesting different types of fats on cholesterol (Ito et. al., 1993),("Types of Fat," 2018).

Trans fatty-acids -Lowers good cholesterol (HDL)
-Raises bad cholesterol (LDL)

Saturated fats -Raises good cholesterol (HDL)
-Raises bad cholesterol (LDL)

Monounsaturated fats -No effect on good cholesterol (HDL), but ratio of good to bad improves
-Lowers bad cholesterol (LDL)

Polyunsaturated fats -Slightly lowers good cholesterol (HDL)
-Lowers bad cholesterol (LDL) - more than monounsaturated fats, so ratio of good to bad improves even more than mono.
-Increased omega-6 fatty acids (a positive change)

It also replaces much of the oil (and hence fats) with applesauce, lowering the overall fat and saturated fat content. This lowers the calorie count, which can be helpful for some patients.

Next, dried cranberries replace the traditional high phosphorus walnuts along with lowering the potassium levels. There are high levels of vitamin A in this cake, so please talk to your care team before making this an everyday dessert.

Ingredients:

For the Cake:

- 2 cups flour
- 2 cup sugar
- 1/2 cup vegetable oil
- 2 1/2 cups shredded carrots
- 4 eggs
- 1 cup dried cranberries
- 2 teaspoons baking soda
- 2 teaspoons cinnamon
- 1 teaspoon allspice
- 1/8 teaspoon salt
- 1 cup apple sauce

For the icing:

- 3 cups powdered sugar
- 12 ounces cream cheese, room temperature
- 2 tablespoons heavy cream
- 1 tablespoon vanilla

Time: 2 hours Yield: 1 two layer cake Portions: 12 Serving Size: 1 piece

Prep 2 nine inch round cake pans by spraying with a non stick spray or using a little butter. I like to insert a wax paper cut out to match the bottom of each pan after applying the spray or butter, then dust with flour.

Preheat oven to 350 degrees.

For the batter use 2 separate bowls, one for the wet and one for dry ingredients. Remember, sugar is always used in the wet mix.

In one bowl combine the flour, baking soda, cinnamon, salt, allspice, and mix well. In the other bowl, whisk the eggs, add the sugar, and mix well. Now add the oil and apple sauce and mix very well. The consistency of the mixture should look and feel a little thick.

Now slowly combine the dry and wet together. When they have come together into a batter, add the carrots and dried cranberries. Split the batter between the 2 round cake pans and bake for 30-40 minutes or until a toothpick comes out dry when inserted in the cake.

Let the cakes rest for 15 minutes. This will give them time to set up and finish cooking. When done, remove and transfer to a cooling rack.

For the frosting, combine the cream cheese, powdered sugar, and vanilla in a mixing bowl and mix until the frosting is smooth and creamy.

When the cakes have completely cooled, it's time to assemble. Place one cake bottom side down on the cake stand and frost the top. I prefer to use a long serrated edge knife to cut of the top arc of the cake, but you can also even it out with frosting. Next place the second cake on the frosted top of the first. Frost the top and edges of the cake as desired.

Making a whole cake may seem like too much work and too much cake. However, I think you will find the end results are worth the effort. Especially because the cake can be frozen. I cut the pieces into single servings, wrap in wax paper and then place individually in a zip lock bag. Into the freezer they go, and they are ready to eat within 20 minutes after removing from the freezer. A perfect treat for the "brain fog" and exhausting days if you are on dialysis.

Carrot Cake

		Fat (g)	Sat. Fat (g)	Phos (mg)	Pot (mg)	Sodium (mg)	Calories	Carbs (g)	Vit A (iu)	Vit C (mg)	Vit K (Mcg)	Vit E (mg) AT	Vit B6 (mg)
	RDA	**65**	**22**	**700**	**3300**	**2300**	**2000**	**275**	**5000**	**60**	**120**	**20**	**2**
Ingredients	**Amt**												
Flour, all purpose	**2 cups**	2.4	0.4	270	268	5	910	190	5	0	0.8	0.3	0.2
		3.69%	1.82%	38.57%	8.12%	0.22%	45.50%	69.09%	0.10%	0.00%	0.67%	1.50%	10.00%
Sugar, granulated	**2 cups**	0	0	0	0	0	1549		0	0	0	0	0
		0.00%	0.00%	0.00%	0.00%	0.00%	77.45%	0.00%	0.00%	0.00%	0.00%	0.00%	0.00%
Sugar, powdered	**3 cups**	0	0	0	7.2	3.6	1401	360	0	0	0	0	0
		0.00%	0.00%	0.00%	0.22%	0.16%	70.05%	130.91%	0.00%	0.00%	0.00%	0.00%	0.00%
Vegetable oil	**1/2 cup**	72.67	5.33	0.00	0.00	0.00	318.33		0.00	0.00	51.67	12.70	0.00
		111.79%	24.24%	0.00%	0.00%	0.00%	15.92%	0.00%	0.00%	0.00%	43.06%	63.50%	0.00%
Carrots	**2 1/2 cup**	0.00	0.00	112.00	880.00	220.75	156.00	30.00	45940.00	19.00	42.25	2.40	0.60
		0.00%	0.00%	16.00%	26.67%	9.60%	7.80%	10.91%	918.80%	31.67%	35.21%	12.00%	30.00%
Eggs	**4 whole**	20	8	382	268	280	284	0	976	0	0.4	2	0
		30.77%	36.36%	54.57%	8.12%	12.17%	14.20%	0.00%	19.52%	0.00%	0.33%	10.00%	0.00%
Heavy Cream	**2 Tbsps**	11.00	6.88	18.50	22.25	11.30	103.00	0.88	437.38	0.18	0.95	0.31	0.01
		16.92%	31.25%	2.64%	0.67%	0.49%	5.15%	0.32%	8.75%	0.29%	0.79%	1.56%	0.63%
Craisins	**1 cup**	3	0	9.6	48	3.6	369	99	0	0.3	4.5	1.2	0
		4.62%	0.00%	1.37%	1.45%	0.16%	18.45%	36.00%	0.00%	0.50%	3.75%	6.00%	0.00%
Baking Soda	**2 tsps**	0	0	0	0	2462	0	0	0	0	0	0	0
		0.00%	0.00%	0.00%	0.00%	107.04%	0.00%	0.00%	0.00%	0.00%	0.00%	0.00%	0.00%
Cinnamon	**2 tsps**	0		3.5	22.1	0	12	4	7.6	0.2	1.6	0.14	0
		0.00%	0.00%	0.50%	0.67%	0.00%	0.60%	1.45%	0.15%	0.33%	1.33%	0.70%	0.00%
Salt (Pinch)	**1/8 tsp**					287.5							
		0.00%	0.00%	0.00%	0.00%	12.50%	0.00%	0.00%	0.00%	0.00%	0.00%	0.00%	0.00%
Cream Cheese	**12 oz**	120	60	356.4	463.2	463.2	1152	12	4248	0	12	2.4	0
		184.62%	272.73%	50.91%	14.04%	20.14%	57.60%	4.36%	84.96%	0.00%	10.00%	12.00%	0.00%
Vanilla	**1 Tbsp**	0.00	0.00	0.01	0.19	0.01	0.37	0.00	0.00	0.00	0.00	0.00	0.00
Applesauce	**1 cup**	0.00	0.00	14.80	184.00	4.90	167.00	43.00	14.80	4.20	1.50	0.40	0.10
		0.00%	0.00%	2.11%	5.58%	0.21%	8.35%	15.64%	0.30%	7.00%	1.25%	2.00%	5.00%
Total for Dish		229.07	80.61	1166.81	2162.94	3741.86	6421.70	738.88	51628.78	23.88	115.67	21.85	0.91
		352.41%	366.40%	166.69%	65.54%	162.69%	321.09%	268.68%	1032.58%	39.79%	96.39%	109.26%	45.63%
# of Servings		12	12	12	12	12	12	12	12	12	12	12	12
Totals per serving		19.09	6.72	97.23	180.25	311.82	535.14	61.57	4302.40	1.99	9.64	1.82	0.08
		29.37%	30.53%	13.89%	5.46%	13.56%	26.76%	22.39%	86.05%	3.32%	8.03%	9.11%	3.80%

Essential Amino Acid Chart for recipe Carrot Cake

	Mg/ gram protein needed	Flour, AP	Recipe 0 g	Heavy Cream	Recipe 6.13 g	Apple Sauce	Recipe .4 g	Cream Cheese	Recipe 28.39 g	Carrots Peeled into strips	Recipe 2.4 g	Total Grams	Total Amino Acid	Total Protein/g
Essential Amino Acids		Protein/g	Recipe	Protein/g	Recipe	Protein/g	Recipe	Protein/g	Recipe	Protein/g	Recipe			
Histidine	18	22.33	13.40	27.14	166.25	18.50	7.40	28.82	818.20	21.67	52.00	37.92	1057.25	27.88
Isoleuciine	25	34.57	20.74	60.20	368.75	43.00	17.20	53.35	1514.61	56.67	136.00		2057.30	54.26
Methionine	25	17.75	10.65	24.69	151.25	12.25	4.90	31.47	893.43	10.00	24.00		1084.23	28.60
Leucine	55	68.84	41.30	97.55	597.50	67.75	27.10	108.24	3072.93	58.33	140.00		3878.84	102.30
Lysine	51	22.09	13.26	79.18	485.00	67.75	27.10	93.53	2655.32	55.00	132.00		3312.67	87.37
Phenylalanine	47	50.39	30.23	48.16	295.00	30.75	12.30	47.94	1361.02	43.33	104.00		1802.55	47.54
Threonine	27	27.21	16.33	45.10	276.25	33.00	13.20	38.35	1088.76	51.67	124.00		1518.53	40.05
Tryptophan	7	12.33	7.40	14.08	86.25	12.25	4.90	11.35	322.23	15.00	36.00		456.77	12.05
Valine	32	40.23	24.14	66.53	407.50	49.25	19.70	65.29	1853.58	60.00	144.00		2448.92	64.59
Amount per recipe		2 cups		2 Tbsps		1 cup		12 oz		3 cup				
Protein per recipe (g)		0.6		6.13		0.4		28.39		2.4				

Total recipe Protein (g)	# of servings	Protein / serving (g)	Not a Complete Protein
37.92	12	3.16	

Pasta with Squash and Carrots in a Cream Sauce

This dish is based on the very popular food trend of veggie spirals. I do not use a "spiralizer" in this recipe, but feel free to break yours out if you have one lying around. You can also buy pre-spiraled packages of vegetables that you can find in the produce department.

I prefer to use my peeler to make strips and use similar shaped pasta like pappardelle or linguine. Having similar shapes will allow you to sneak in some veggies in a dish for those who might be averse.

Food Science Note

As the name suggests we are going to make a pasta using a cream sauce. If you are at all familiar with the CKD diet you should be asking yourself how is this acceptable. It is well known that milk and dairy products contain high levels of phosphorus and should be limited because of the potential long term problems associated with high phosphorus levels. To make things worse we are going to use heavy cream, a product with high fat content.

Milk comes from the cow as whole milk, which is about 3.5% milk fat. The rest is water, proteins, and sugars. By removing some of the fat we end up with the lower fat milk products like 2% and skim.
If we remove the water content and other ingredients that it contains, we get the higher fat content items like half & half and heavy cream. When the other ingredients are removed with the water, we also remove many of the proteins, which contain much of the phosphorus (McGhee, 1984, p 9).

The higher fat content in heavy cream can frighten consumers and dietitians alike. However, the amount of heavy cream per portion keeps the levels reasonable. The other advantage is that fat absorbs other flavors that are cooked in it. Increasing the overall flavor in the dish will reduce the need to add salt. The sodium in salt causes the body to retain fluid and can cause many difficulties for CKD patients.

The ease and speed of this recipe allows for you to make a quick meal that will look and taste like a chef stopped by the house to cook for your family. In addition, the kidney friendly nature of this recipe will not deter those at the table who do not have those difficulties because it is loaded with flavor.

Ingredients:

- 1 pound linguine or pappardelle or other wide pasta
- 1 zucchini
- 1 yellow squash
- 1 cup heavy cream
- 1 medium sized carrot
- 3 cloves garlic, minced
- 8 leaves fresh basil, chopped
- 2 tablespoons butter
- 1/2 cup freshly grated Parmigiana Reggiano
- 1/8 teaspoon salt

Time: 20 minutes Yield: 36 oz. Portions: 8
Serving size: 4.5 oz

Start by placing a large pasta pot with 4 to 6 quarts of water on the stove and bring to a boil. This dish will not take long, so once you start cooking your pasta it will be done soon after the pasta is ready.

Peel your squash and zucchini in wide strips until you see the seeds. I like to use the skin, so I clean the veggies well before peeling. Remove the outer layer of the carrot and discard, then peel the carrot in the same way as the squash and zucchini. Set all the veggies on one plate.

Heat a large skillet that will fit all the pasta on medium heat and melt the butter. As soon as the butter is melted add the minced garlic and sauté for just a few minutes. Garlic is sensitive, and if it burns, all the fat that it is cooking in will be ruined. Burnt garlic cannot be saved, nor the fat it was overcooked in.

If you are using dried pasta if will take about 8-11 minutes to cook, about the same time it will take to finish the sauce. If you are using fresh pasta, it will only take a few minutes, so time your pasta accordingly.

Once the garlic has begun to create an aroma in the kitchen, it's time to add the heavy cream. You will want to let the cream reduce about 1/3 to thicken up the sauce. About halfway through the thickening process, add the basil to flavor the cream with the herb.

If you have a steaming basket that will fit on top of the pasta water, place the the squash, zucchini, and carrot strips in the basket. Once you add the pasta in the water, put the veggie basket on top of the water and steam veggies while the pasta cooks.

The steaming of the veggies will retain more of the vitamins and minerals. Alternatively, cook the veggies in the cream sauce while the pasta cooks. I like my veggies a little crisp, but you can certainly cook them longer.

Once the pasta is nearly done (al dente), remove the veggies from the steamer and place directly in the sauce. This will add more flavor to the sauce.

Next, remove the pasta from the water and add directly to the sauce to finish cooking, stirring thoroughly to mix the pasta, cream sauce and veggies.

Food Note

Fats attract flavors, slowly when cool, and quickly when heated. In this case the high concentration of fat in the heavy cream is a sponge for the flavors of the garlic, basil, and other veggies. Pasta absorbs liquid to rehydrate. This is why we salt the water as it flavors the pasta. By finishing the cooking process in a sauce loaded with the flavors of the rest of the dish, the pasta absorbs those flavors as well. The starchy water left on the pasta also thickens the sauce.

When the pasta is fully cooked, turn off the heat and add the Parmigiano Reggiano, mix and serve.

In the past we have been guided to restrict the fat content in our cooking for better health. This long standing thought is being challenged by several new studies (Rippe & Angelopoulos, n.d.). Although those with high levels of cholesterol are still considered "at risk" when it comes to the consumption of fats in their diet. Check with your doctor and nutritionist.

Phosphorous in heavy cream is significantly lower that in all other milk products and because of the higher fat, and thus higher calorie count, you will tend to eat a smaller portion to feel full, keeping the phosphorous levels lower.

Pasta w/ Squash, Carrots in a Cream Sauce

		Fat (g)	Sat. Fat (g)	Phos (mg)	Pot (mg)	Sodium (mg)	Calories	Carbs (g)	Vit A (iu)	Vit C (mg)	Vit K (Mcg)	Vit E (mg) AT	Vit B6 (mg)
	RDA	**65**	**22**	**700**	**3300**	**2300**	**2000**	**275**	**5000**	**60**	**120**	**20**	**2**
Ingredients	**Amt**												
Pasta	**1 lb**	1	0	182.4	86.4	337.6	1600	104	0	0	0	0	0
		1.54%	0.00%	26.06%	2.62%	14.68%	80.00%	37.82%	0.00%	0.00%	0.00%	0.00%	0.00%
Squash - Yellow	**1 Med**	0	0	74.5	514	4	31	8	392	33.3	5.9	0	0.4
		0.00%	0.00%	10.64%	15.58%	0.17%	1.55%	2.91%	7.84%	64.80%	4.92%	0.00%	20.00%
Zucchini	**1 Med**	0.00		74.50	514.00	4.00	31.00	8.00	392.00	33.30	5.90	0.00	0.40
		0.00%	0.00%	10.64%	15.58%	0.17%	1.55%	2.91%	7.84%	55.50%	4.92%	0.00%	20.00%
Carrots - Peeled	**1 Med**	0.00	0.00	21.40	195.00	42.00	25.00	12.00	10190.	3.60	8.10	0.40	0.10
(61 g)		0.00%	0.00%	3.06%	5.91%	1.83%	1.25%	4.36%	203.80	6.00%	6.75%	2.00%	5.00%
Parmigiano Reggiano	**1 oz**	7	5	172	13.2	428	121	1	112	0	0.5	1.2	0.3
		10.77%	22.73%	24.57%	0.40%	18.61%	6.05%	0.36%	2.24%	0.00%	0.42%	6.00%	15.00%
Heavy Cream	**1 Cup**	44.00	55.00	74.00	89.00	45.00	410.00	7.00	1750.0	0.70	3.80	1.25	0.10
		67.69%	250.00%	10.57%	2.70%	1.96%	20.50%	2.55%	35.00%	1.17%	3.17%	6.25%	5.00%
Butter	**2 Tbsp**	22	14	0	1	1	442	0	0	0	30.1	7.2	0
		33.85%	63.64%	0.00%	0.03%	0.04%	22.10%	0.00%	0.00%	0.00%	25.08%	36.00%	0.00%
Fresh Basil	**1/4 cup**	0	0	18.1	77	7.6	9	0	1179.5	18.6	230	0.1	0
		0.00%	0.00%	2.59%	2.33%	0.33%	0.45%	0.00%	23.59%	31.00%	191.67%	0.50%	0.00%
Garlic - Fresh	**2 cloves**	0.00	0.00	23.20	117.00	1.00	4.00	2.00	0.27	0.90	0.07	0.00	0.00
		0.00%	0.00%	3.31%	3.55%	0.04%	0.20%	0.73%	0.01%	1.50%	0.06%	0.00%	0.00%
Salt (Pinch)	**1/8 tsp**					287.5							
		0.00%	0.00%	0.00%	0.00%	12.50%	0.00%	0.00%	0.00%	0.00%	0.00%	0.00%	0.00%
Pepper	**1/4 tsp**	0.00	0.00	2.85	7.58	0.18	2.28	0.50	1.50	0.10	0.83	0.00	0.00
		0.00%	0.00%	0.41%	0.23%	0.01%	0.11%	0.18%	0.03%	0.17%	0.69%	0.00%	0.00%
Total for Dish		74.00	74.00	642.95	1614.18	1157.88	2675.28	142.50	14017.	90.50	285.19	10.15	1.30
		113.85	336.36%	91.85%	48.91%	50.34%	133.76%	51.82%	280.35	150.83	237.66%	50.75%	65.00%
# of Servings		8	8	8	8	8	8	8	8	8	8	8	8
Totals per serving		9.25	9.25	80.37	201.77	144.73	334.41	17.81	1752.1	11.31	35.65	1.27	0.16
		14.23%	42.05%	11.48%	6.11%	6.29%	16.72%	6.48%	35.04%	18.85%	29.71%	6.34%	8.13%

Essential Amino Acid Chart for Pasta with Squash in a Cream Sauce

	Mg/g Prote.	Sq.	Rec. 6.1 g	Pasta	Rec. 56 g	Parm Regg	Rec. 10 g	Basil	Rec. .9 g	H. Cr.	Rec. 4.9 g	But.	Rec. .4 g	Fresh Garlic	Rec. .9 g	Carr.	Rec. .8 g	Tot. g	Tot. Am. Acid	Total P/G
Ess. Amino Acids		**P/G**	**Rec.**	**P/G**	**Rec.**	**P/G**	**Rec.**	**P/G**	**Rec.**	**P/G**	**Rec.**	**P/G**	**Rec.**	**P/G**	**Rec.**	**P/G**	**Rec.**			
Histidine	**18**	21.67	132.19	20.28	1135.6	38.70	387.00	15.89	14.30	27.14	133.00	27.47	5.49	17.00	3.40	21.67	17.33	79.1	1663.7	21.03
Isoleuciine	**25**	36.40	222.04	38.52	2157.1	53.00	530.00	32.33	29.10	60.20	295.00	61.05	12.21	32.50	6.50	56.67	45.33		2927.8	37.01
Methionine	**25**	14.87	90.73	15.52	869.12	26.80	268.00	11.22	10.10	24.69	121.00	25.11	5.02	11.50	2.30	10.00	8.00		1235.1	15.62
Leucine	**55**	58.76	358.45	68.00	3808.0	96.70	967.00	59.44	53.50	97.55	478.00	98.95	19.79	47.00	9.40	58.33	46.67		5162.6	65.27
Lysine	**51**	55.43	338.11	19.16	1072.9	92.58	925.84	34.22	30.80	79.18	388.00	80.00	16.00	42.00	8.40	55.00	44.00		2361.3	29.85
Phenylalanine	**47**	35.55	216.86	48.40	2710.4	53.80	538.00	40.44	36.40	48.16	236.00	49.00	9.80	27.50	5.50	43.33	34.67		3480.5	44.00
Threonine	**27**	24.01	146.47	26.44	1480.6	36.90	369.00	32.33	29.10	45.10	221.00	45.42	9.08	23.50	4.70	51.67	41.33		2009.9	25.41
Tryptophan	**7**	8.27	50.45	12.76	714.56	13.50	135.00	12.11	10.90	14.08	69.00	14.32	2.86	10.00	2.00	15.00	12.00		904.88	11.44
Valine	**32**	44.69	272.60	42.40	2374.4	68.70	687.00	39.56	35.60	66.53	326.00	67.89	13.58	43.50	8.70	60.00	48.00		3356.2	42.43
Amount per recipe		3 each		1 lb		1 oz		1 oz.		1 cup		2 tbsp		1 clove		1 cup				
Protein per recipe (g)		6.1		56		10		0.9		4.9		0.2		0.2		0.8				

Total recipe Protein (g)	**# of servings**	**Protein / serving (g)**	Not a Complete Protein
79.1	8	9.8875	

*P/G is protein per gram.

Texas Chili or Bowl o'Red

Chili is another food item that CKD patients are told to limit or even eliminate from their diets. There a several factors that make this meal a difficult one for those with kidney issues.

Let's start with the idea of chili. Depending on where you live, the idea of a hot bowl of chili conjures up different pictures for different people. For most it is a spicy mix of tomatoes, meat, beans and chili peppers. It's a thick stew-like meal that leaves you warm all over.

However, the many different varieties of chili have different health issues for CDK patients. Beef, chicken, veggie, white, Cincinnati, Texas, California, and others.

The common ingredients of chili - tomatoes, beans, meat, spice, and often cheese -create a legitimate cause for concern for CKD patients and their dietitians. Tomatoes are high in potassium. Beans, meat, and cheese are all high phosphorus foods, and have high levels of protein. When served together they can be a real problem for those with phosphorus issues.

This recipe is going to address a few of those issues by choosing what is arguably called the original chili recipe, Texas Chili or a "Bowl o'Red." In this version, the stew is made without tomatoes and without beans. It uses beef, usually a cut that takes a few hours of cooking to attain a tender and delicious state. There will still be a higher level of phosphorus than may be acceptable for some patients, however, there is also a high level of protein.

For those not familiar with this Texas delicacy (I was included in that group), you may be wondering how the meal turns out red without the tomatoes. Well, I am going to tell you, if the true Texans will forgive my intrusion into their cuisine.

The key to the color is using dried chili peppers. The peppers are rehydrated in water, beef, or chicken stock, along with other flavors making the final dish red. There are several different types of dried chili peppers and each has a different flavor profile. Sweet, spicy, fruity, and smokey are some of the profiles for a chili pepper. Just like grapes or coffee beans, your pepper choices will bring out certain flavors in the final dish. In this recipe I use a combination of peppers to get a complex and full bodied chili.

Ingredients:

- 8 ounces dried chili peppers with stems and seeds removed.
  - 3 Sweeter - Costeno, Guajillo, or New Mexico
  - 3 Fruity- Ancho or Pasilla are easy to find.
  - 3 Spicy (more if you like it spicier) - Arbol or Cascabel
- 2 1/2 pounds beef chuck. I find it best to buy a large piece and cut it into 3/4 inch cubes
- 2 cups beef or chicken stock - homemade is preferred, low sodium otherwise
- 2 cups water, more if needed
- 1/2 cup chopped onion
- 2 chipotle peppers in adobo sauce
- 3 cloves fresh garlic
- 1 tablespoon brown sugar
- 1/4 cup vegetable oil
- 2 tablespoons apple cider or white wine vinegar
- 1/2 cup masa harina, corn flour used to make tortillas. It is used here as a thickener (you can replace masa with 1 or 2 corn tortillas as needed).
- 1/8 teaspoon salt
- 1 tablespoon of ground cumin
- 1/2 teaspoon ground allspice
- Black pepper to taste
- Lime wedges for garnish

Time: 2 1/2 hours Yield: 64 oz Portions: 12
Serving Size: 6 oz. (3 oz. meat, 3 oz. of sauce)

Let's break this down to a few different sections, each ending as a part of the recipe that we will combine together and stew for 2 hours. The first step is to make the pepper paste. This is the rehydrated pepper mix. The second part is to build the meat mixture. The final part is to combine the two and simmer.

Step 1 - Pepper Paste. In one pot, place one cup stock and one cup water on medium heat. Toast dried peppers by placing a separate large pot for the chili on medium heat and toss the dried chiles into the pot for about a minute. Flip the peppers once while toasting, and when done remove and set them in the hot liquid and turn off the heat. Let the peppers and liquid sit for at least 10 -15 minutes and cool down (Hot fluid in the blender is a disaster waiting for you to push the "on" button).

Now place the peppers and half the soaking fluid in a blender, add salt, pepper, cumin, allspice, and chipotle peppers. Blend well until a paste forms. This is the base for the chili. You may want to add some more of the soaking fluid to help make a smoother paste.

Step 2. Build the meat mixture. Using the large pot heat the oil on medium to high heat and add your beef chuck pieces in 2 to 3 batches, making sure the pieces are in a single layer. Each piece needs to come in contact with the bottom of the pot. Rotate the beef at least once to brown on all sides. Remove the first batch and continue until all meat is brown and you have a pile set aside. Now add the onion to the pot with all the bits left in it, and sauté for 3-5 minutes. Add the garlic and continue cooking for another minute.

Step 3. Add the meat back into the pot, along with the rest of the stock and pepper paste. Stir thoroughly and add, vinegar, masa harina and 1/2 of the remaining water. Bring the mixture to a boil and reduce to a simmer.

Add the rest of the water as needed if the mixture gets too thick. Cook this with a cover slightly ajar for 2 hours, stirring occasionally. This will give the meat time to tenderize as the collagen in the meat disintegrates into the chili.

Food Note

You could use chicken in the recipe and if you do, I suggest chicken thighs. Cook everything as above except do not add the browned chicken until the last 45 minutes.

White Chili is made with a base of chicken stock, chicken, white beans, and usually green peppers

Cincinnati chili has a tomato and ground beef base, and is served over spaghetti along with other layering options. You order it as chili three way, four way, or five way. Layer one is the spaghetti. Layer two is the tomato/beef mixture. Layer three is the cheese. Chili three way includes these first layers. Chili 4 way adds beans or onions and 5 way adds both beans and onions.

Texas Chili (Bowl o' Red)

		Total Fat (g)	Sat. Fat (g)	Phos (mg)	Pot (mg)	Sodium (mg)	Calories	Carbs (g)	Vit A (iu)	Vit C (mg)	Vit K (Mcg)	Vit E (mg) AT	Vit B6 (mg)
	RDA	65	22	700	3300	2300	2000	275	5000	60	120	20	2
Ingredients	Amt												
Dried Chilis	9 total												
Total of 2 oz		2	0	89	1048	51	182	40	14834	17.6	60.6	1.8	0.4
		3.08%	0.00%	12.71%	31.76%	2.22%	9.10%	14.55%	296.68	29.33%	50.50%	9.00%	20.00%
Beef Chuck	2 1/2 lbs.	48.9	18.9	1760.1	1840.8	468	1660	0	0	0	12	3.0	3
		75.23%	85.91%	251.44%	55.78%	20.35%	83.00%	0.00%	0.00%	0.00%	10.00%	15.00%	150.00%
Beef Stock	2 cups	0	0	149.2	888	950	62	6	0	0	0.4	0	0.2
		0.00%	0.00%	21.31%	26.91%	41.30%	3.10%	2.18%	0.00%	0.00%	0.33%	0.00%	10.00%
Onion	1/2 cup	0	0	23.2	117	3.2	32	0	1.6	5.9	0.3	0	0.1
		0.00%	0.00%	3.31%	3.55%	0.14%	1.60%	0.00%	0.03%	9.83%	0.25%	0.00%	5.00%
Chipotle Pepper	2 plus liquid	0	0	8	85	736	12	2	748	4.4	5.8	0.4	0
		0.00%	0.00%	1.14%	2.58%	32.00%	0.60%	0.73%	14.96%	7.33%	4.83%	2.00%	0.00%
Garlic - Fresh	3 Cloves	0	0	13.8	36.1	1.5	13	3	0.8	2.8	0.2	0	0.1
		0.00%	0.00%	1.97%	1.09%	0.07%	0.65%	1.09%	0.02%	4.67%	0.17%	0.00%	5.00%
Brown Sugar	1 Tbsp	0.00	0.00	0.00	0.00	0.00	48.38	12.50	0.00	0.00	0.00	0.00	0.00
		0.00%	0.00%	0.00%	0.00%	0.00%	2.42%	4.55%	0.00%	0.00%	0.00%	0.00%	0.00%
Lard or Veg. oil	1/4 cup	54.50	4.00	0.00	0.00	0.00	481.75	0.00	0.00	0.00	38.75	9.53	0.00
		83.85%	18.18%	0.00%	0.00%	0.00%	24.09%	0.00%	0.00%	0.00%	32.29%	47.63%	0.00%
White Wine Vinegar	2 Tbps.	0.00	0.00	2.39	11.65	2.39	5.63	0.00	0.00	0.15	0.00	0.00	0.00
		0.00%	0.00%	0.34%	0.35%	0.10%	0.28%	0.00%	0.00%	0.25%	0.00%	0.00%	0.00%
Masa Harina	1/2 cup	2	0.5	127	170	2.85	208	43.5	122	0	0	0	0.2
		3.08%	2.27%	18.14%	5.15%	0.12%	10.40%	15.82%	2.44%	0.00%	0.00%	0.00%	10.00%
Salt (Pinch)	1/8 tsp					287.50							
						12.50%							
Total for Dish		107.40	23.40	2172.69	4196.55	2502.44	2704.75	107.00	15706.	30.85	118.05	14.73	4.00
		165.23%	106.36%	310.38%	127.17%	108.80%	135.24%	38.91%	314.13	51.42%	98.38%	73.63%	200.00%
# of Servings		12	12	12	12	12	12	12	12	12	12	12	12
Totals per serving		8.95	1.95	181.06	349.71	208.54	225.40	8.92	1308.8	2.57	9.84	1.23	0.33
		13.77%	8.86%	25.87%	10.60%	9.07%	11.27%	3.24%	26.18%	4.28%	8.20%	6.14%	16.67%

Essential Amino Acid Chart for recipe Texas Chili

	Mg/g Prot.	Pep. Dried Chili	Recipe 6 g	Beef, Chuck	Recipe 324 g	Beef Stock	Recipe 9.4 g	Onion	Recipe .9 g	Masa	Recipe .2 g	Chip. Pepp.	Recipe .4 g	Tot. Gr.	Total Amino Acid	Total P/G
Ess. Am. Acids		**P/G**	**Total**	**P/G**	**Total**	**P/G**	**Total**	**P/G**	**Total**	**P?g**	**Total**	**P/G**	**Total**			
Histidine	**18**	20.70	124.20	31.98	8442.72			12.44	11.20	27.47	5.49	20.00	8.00	280.9	8580.41	30.55
Isoleuciine	**25**	31.93	191.58	45.56	12027.84			11.79	10.61	61.05	12.21	33.00	13.20		12244.8:	43.59
Methionine	**25**	11.87	71.22	26.05	6877.20			1.68	1.52	25.11	5.02	13.00	5.20		6958.64	24.77
Leucine	**55**	51.67	310.02	79.63	21022.32			21.05	18.95	98.95	19.79	53.00	21.20		21373.3:	76.09
Lysine	**51**	44.00	264.00	84.57	22326.48			32.84	29.56	80.00	16.00	46.00	18.40		22624.8	80.54
Phenylalanine	**47**	30.53	183.18	39.51	10430.64			22.22	20.00	49.00	9.80	32.00	12.80		10636.4:	37.87
Threonine	**27**	36.33	217.98	40.00	10560.00			17.68	15.92	45.42	9.08	36.50	14.60		10801.6	38.45
Tryptophan	**7**	12.60	75.60	6.57	1734.48			12.44	11.20	14.32	2.86	13.00	5.20		1818.14	6.47
Valine	**32**	41.67	250.02	49.63	13102.32			18.67	16.80	67.89	13.58	43.00	17.20		13383.1:	47.64
									0.00							
Amount per recipe		2 oz.		2 1/2 lb 30 oz.		2 Cup		1/2 cup		2 tbsp		2 oz.				
Protein per recipe (g)		6		264		9.4		0.9		0.2		0.4				

Total recipe Protein (g)	# of servings	Protein /serving (g)	Complete Protien
280.9	12	23.41	

Three Sisters (with and without rice)

This dish is made of winter squash, corn, and climbing beans. The combination was grown together and provided plenty of sustenance to the Native American people. This group of veggies has the added benefit of being one of the few vegetable combinations that create a complete protein (containing enough of the 9 essential amino acids per gram of protein). The dish does not provide high levels of protein, but it is considered a complete protein which can be beneficial for different stages of CKD.

This dish can be served as a side or with a starch (rice or pasta) making it a main course. Below I have given you the recipe with rice. In addition, the nutritional data charts show the per serving amounts with and without rice.

Ingredients:

- 1 1/2 cups zucchini, cut into thick pieces (cubes, rounds, sticks)
- 1 1/2 cups yellow squash (same as above)
- 2 cups green snap beans (these are immature beans)
- 1 cup corn kernels - frozen or fresh
- 2 cups cooked rice (optional)
- 1/2 cup onion, chopped
- 1 clove fresh garlic
- 1 tablespoon fresh lime juice
- 2 tablespoons olive oil
- 1 tablespoon cumin
- 1/4 teaspoon salt
- Fresh ground pepper

Total Time: 40 minutes Yield: 6 cups Portions: 4
Serving size: 1 1/2 cups

In this recipe I am going to use 2 different cooking methods for the vegetables, roasting and steaming. The first method is quicker and easier.

Mix all the veggies in a bowl with the oil, salt, and pepper. Place on a sheet pan and roast at 400 degrees for 20-30 minutes and serve.

Food Note

The three sisters refers to crops that were grown together by the Native American people dating back to several centuries before the Europeans starting traveling to the continent. Corn, winter squash, and climbing beans make up this dish and also the crops that are grown together, using benefits of each to aid the other. The corn stalk is used as a platform for the beans and as a base to climb. The squash covers the ground to inhibit the growth of weeds and other vegetation, leaving the vital nutrients for the three sisters. The squash also uses its prickly leaves to keep animals away. The beans supply nitrogen to the dirt for all three to thrive. This symbiotic relationship also provides a compete protein when eaten together.

Green snap beans can become dull in color and less appealing when roasted. The other option is to steam the beans, and roast the veggies. Steaming retains as much of the vitamin content in veggies as possible and boiling loses many vitamins to the water. They can also be microwaved, but it is important to add some water to the beans in order for the science of the microwave to work properly, keeping in mind this will also take away some of the nutrients.

Start by preheating the oven to 400 degrees F.
Take the frozen corn and rinse under cool water and place on a paper towel to dry. Now rinse and cut the squash into 1/2" rings, 1" cubes, or sticks. Just make sure they are thick and large because squash will cook quickly and can become mushy. I prefer larger pieces of squash so they hold their form better when cooked.

Next, place a pot with enough water to reach the bottom of the steaming basket on high heat. Add your beans to the basket and mix occasionally.

Add your corn, squash, and chopped onion to a bowl and mix with olive oil, salt and pepper. Mix thoroughly and place evenly on a sheet pan. Cook for 20 minutes in the oven.

I usually start my rice around the same time as the roasting, as it will take about 20-30 minutes total to cook.

When all the ingredients are cooked, combine the veggies and add lime juice and cumin. Mix with the rice, if using, and serve. You could also serve the veggies and the rice separately.

Health Note

Food with higher amounts of protein always comes with high levels of phosphorus. High phosphorus can be very dangerous for CKD patients, especially those on dialysis where the kidneys are not filtering out the excess. But the body still needs the amino acids that make up protein. Finding foods and recipes that can provide high levels of protein and low levels of phosphorus is a challenge. This is called the phosphorus to protein ratio and the lower the number, the better.
Beans are well known as a good source of protein but also have high levels of phosphorus. When mixed with rice (or some other types of starches) they can become complete proteins.
In this recipe, I mentioned using immature beans, those early in the growth cycle. Green snap beans and edamame (immature soy beans) have significantly lower levels of phosphorus without lowering the protein content. Both have a lower phosphorus/protein ratio.

Three sisters with Rice

		Fat (g)	Sat. Fat (g)	Phos (mg)	Pot (mg)	Sodium (mg)	Calories	Carbs (g)	Vit A (iu)	Vit C (mg)	Vit K (Mcg)	Vit E (mg) AT	Vit. B6 (mg)
	RDA	65	22	700	3300	2300	2000	275	5000	60	120	20	2
Ingredients	Amt												
Squash - Both	3 cups	0	0	182.4	888	6.9	54	12	678	57.6	10.2	0.3	0.9
		0.00%	0.00%	26.06%	26.91%	0.30%	2.70%	4.36%	13.56%	96.00%	8.50%	1.50%	45.00%
Gr. Snap beans immature	2 Cups	0	0	83.6	693	13.2	68	15.6	1518	35.2	31.6	1	0.2
		0.00%	0.00%	11.94%	21.00%	0.57%	3.40%	5.67%	30.36%	58.67%	26.33%	5.00%	10.00%
Corn Kernesl	1 Cup	2		137	416	23.1	132	29	0	10.5	0.5	1.2	0.1
		3.08%	0.00%	19.57%	12.61%	1.00%	6.60%	10.55%	0.00%	17.50%	0.42%	6.00%	5.00%
Rice - cooked	2 cups	0.8	0.1	133.6	107.8	2	484	53.2	0	0	0	0.1	0.1
		1.23%	0.45%	19.09%	3.27%	0.09%	24.20%	19.35%	0.00%	0.00%	0.00%	0.50%	5.00%
Chopped onion	1/2 cup	0	0	32.2	116	3.2	32	0	1.6	5.9	0.3	0	0.2
		0.00%	0.00%	4.60%	3.52%	0.14%	1.60%	0.00%	0.03%	9.83%	0.25%	0.00%	10.00%
Garlic Fresh	1 clove	0	0	4.6	12	0.5	4	1	2.67	0.93	0.65	0	0.03
		0.00%	0.00%	0.66%	0.36%	0.02%	0.20%	0.36%	0.05%	1.56%	0.54%	0.00%	1.50%
Fresh lime juice	1 Tbsp	0	0	4.3	36	0.6	8	1	7	9.2	0.2	0.1	2
		0.00%	0.00%	0.61%	1.09%	0.03%	0.40%	0.36%	0.14%	15.33%	0.17%	0.50%	100.00%
Olive oil	2 Tbsp	28	3.75	0.275	0.5375	0	28	0	0	0	16.2	4	0
		43.08%	17.05%	0.04%	0.02%	0.00%	1.40%	0.00%	0.00%	0.00%	13.50%	20.00%	0.00%
Cumin	1 Tbsp	1	0	29.9	107	10.10	22	3	76.2	0.5	0.3	0.2	0
		1.54%	0.00%	4.27%	3.24%	0.44%	1.10%	1.09%	1.52%	0.83%	0.25%	1.00%	0.00%
Salt	1/4 Tsp					575.00							
						25.00%							
Total for Dish		31.8	3.85	607.875	2376.3375	634.60	832	114.8	2283.47	119.83	59.95	6.9	3.53
		48.92%	17.50%	86.84%	72.01%	27.59%	41.60%	41.75%	45.67%	199.72%	49.96%	34.50%	176.50%
# of Servings		4	4	4	4	4	4	4	4	4	4	4	4
Totals (w/ rice)		7.95	0.96	151.97	594.08	158.65	208.00	28.70	570.87	29.96	14.99	1.73	0.88
		12.23%	4.38%	21.71%	18.00%	6.90%	10.40%	10.44%	11.42%	49.93%	12.49%	8.63%	44.13%
Totals (no rice)		31.00	3.75	474.28	2268.54	632.60	348.00	61.60	2283.47	119.83	59.95	6.80	3.43
		47.69%	17.05%	67.75%	68.74%	27.50%	17.40%	22.40%	45.67%	199.72%	49.96%	34.00%	171.50%
Total per serving (no rice)		7.75	0.94	118.57	567.13	158.15	87.00	15.40	570.87	29.96	14.99	1.70	0.86
		11.92%	4.26%	16.94%	17.19%	6.88%	4.35%	5.60%	11.42%	49.93%	12.49%	8.50%	42.88%

Essential Amino Acid Chart for recipe Three Sisters

	Mg needed per gram of Protein	Squash	Recipe 4.5 g	Green Snap Beans	Recipe 4 g	Corn	Recipe 5 g	White Rice raw	Recipe 8.4 g	On.	Recipe .9 g	Total Gram	Total Amino Acid	Tot. P/g
Essential Amino Acids		p/g	Recipe	p/g	Recipe	p/g	Recipe	p/g	Recipe	p/g	Recipe			
Histidine	18	21.67	97.52	18.7	74.80	27.4	137.00	23.69	199.00	12.44	11.20	22.8	519.52	22.79
Isoleuciine	25	36.40	163.80	36.3	145.20	39.8	199.00	43.57	366.00	11.79	10.61		884.61	38.80
Methionine	25	14.87	66.93	12.1	48.40	20.6	103.00	23.69	199.00	1.68	1.52		418.85	18.37
Leucine	55	58.76	264.43	61.5	246.00	107.2	536.00	83.57	702.00	21.05	18.95		1767.38	77.52
Lysine	51	55.43	249.42	48.4	193.60	42.2	211.00	36.43	306.00	32.84	29.56		989.58	43.40
Phenylalanin e	47	35.55	159.98	36.85	147.40	42.6	213.00	53.57	450.00	22.22	20.00		990.38	43.44
Threonine	27	24.01	108.05	43.45	173.80	39.8	199.00	36.19	304.00	17.68	15.92		800.77	35.12
Tryptophan	7	8.27	37.22	10.45	41.80	7.08	35.40	11.67	98.00	12.44	11.20		223.62	9.81
Valine	32	44.69	201.10	49.5	198.00	57	285.00	61.67	518.00	18.67	16.80		1218.90	53.46
Amount per recipe		3 cups		2 Cups		1 cup		2 Cup		1/2 cup				
Protein per recipe (g)		4.5		4		5		8.4		0.9				

Total recipe Protein (g)	# of servings	Protein / serving (g)	Complete Protein
22.8	4	5.7	

Chile Paste

Chile paste is a core ingredient in Texas Chili, however, it can also be used in many other recipes and as a condiment. Just like other homemade condiments, the cook gets to control the ingredients, and hence the taste and health properties of the final product. The amounts of sodium, potassium, and sugars in many chili pastes can be an issue for CKD patients.

This recipe uses no tomatoes and has low potassium, limited amounts of salt for low sodium levels, and no added sugars. You can add a sweetener if you like, but it isn't necessary for a good end product. I have used this in my BBQ sauce recipe as well as chili, rice dishes, and anything that may call for cayenne pepper.

Chile paste is made from dried chiles, stock or water, and a few aromatics. Dried chiles come in many varieties and by combining the different flavor profiles of each pepper, you can tailor the paste to your liking.

The primary flavor profiles can be categorized as fruity, smokey, spicy, sweet, and acidic. Each type of pepper will have a degree of some, if not all of these flavors.

Arbol, Pequin, Morita are examples of spicier peppers. Guajillo, Cascabel, Passilla, and Ancho are fruitier. Chipotle, Mulato,and Morita (again) have a more smokey flavor. You can see that some of these will overlap as they each have complex flavors of their own, similar to grapes for wine.

Health Note

Capsaicin again. Yes, we are going to talk about more of the properties of this compound. One of the more recent developments, not only in the CKD medical community, but in overall health, is inflammation. At one point it was considered only a symptom of certain medical issues, but now it is also considered a cause. Inflammation is also an issue in patients with CKD and finding ways to reduce it in the body can be beneficial.
That's where peppers come in. Chili peppers (not sweet peppers) have high levels of capsaicin and they have been shown to decrease swelling and inflammation in the body as well as pain (Jolayemi & Ojewole, 2013).

Ingredients:

- 2 ounces dried chiles
  - 3 hot (Arbol, Pequin, or Mortia)
  - 3 fruity (Guajillo, Pasilla, Ancho, or Cascabel)
  - 3 sweeter (Costeno, New Mexico)
- 2 chipotle peppers in adobo sauce
- 2 tablespoons of the adobo sauce
- 1/2 cup onion, chopped
- 2 cloves garlic, minced.
- 1/2 cup stock (chicken, veggie, beef) or water
- 1/2 cup of the reserved water from the pepper soak
- 2 tablespoons of fresh lime juice
- 1/2 tablespoon vegetable oil
- 1 teaspoon ground cumin
- 1 teaspoon salt
- 1 tablespoon brown sugar (optional)

Total time: 30 min. Yield: 1 1/2 cups Portions: 12
Serving size: 2 Tbsp

The first step is to toast the peppers and rehydrate them.

Heat 2 cups of water on medium heat. Place a heavy bottom pan on medium heat and put all the peppers in the pan and watch them carefully. It only takes a few minutes for the peppers to toast. Flip them over, and toast the other side. Don't let the pan smoke or you will burn the peppers and have to start with new ones.

When the peppers are toasted, remove the stems and seeds, and place the remaining parts in the hot water and turn off the heat. Let sit for 15 to 30 minutes. When soft, remove from liquid and reserve 1 cup of the soaking fluid. It should be a nice red color.

While the peppers soak, heat the pan you used to toast the peppers on medium heat to sweat the onions and garlic. Add oil when pan is hot. Start with the onions as they will take longer (3-5 minutes). Add the garlic and sweat for another minute. Remove from heat and set aside.

In a blender or food processor, combine the peppers, garlic/ onion mixture, chipotle, adobo, and lime juice. Add the sugar if using.

Add about 1/3 of the stock (or water) and the 2 tablespoons of the reserved red pepper water and blend. At this point you should have a thick paste, and you can keep adding liquid until you get a consistency of your choice.

A thick paste has greater options as it can be mixed with other liquids without watering down a recipe. A medium thickness is good to add directly to food, or in a recipe. I stay away from adding too much liquid because I don't want to water down the flavor.

Place in a mason jar, cover and refrigerate. This will remain fresh for up to 3 weeks.

Chile Paste

		Total Fat (g)	Sat. Fat (g)	Phos (mg)	Pot (mg)	Sodium (mg)	Calories	Carbs (g)	Vit A (iu)	Vit C (mg)	Vit K (Mcg)	Vit E (mg) AT	Vit B6 (mg)
	RDA	65	22	700	3300	2300	2000	275	5000	60	120	20	2
Ingredients	Amt												
Dried Chilis													
Pasilla Chili	3												
Guajillo	3												
Arbol	3												
Total of 2 oz		2	0	89	1048	51	182	40	14834	17.6	60.6	1.8	0.4
		3.08%	0.00%	12.71%	31.76%	2.22%	9.10%	14.55%	296.68	29.33%	50.50%	9.00%	20.00%
Chicken/Beef Stock	1/2 cup	0	0	37.30	222.00	237.50	15.50	1.50	0.00	0.00	0.10	0.00	0.05
		0.00%	0.00%	5.33%	6.73%	10.33%	0.78%	0.55%	0.00%	0.00%	0.08%	0.00%	2.50%
Onion	1/2 cup	0	0	23.2	117	3.2	32	0	1.6	5.9	0.3	0	0.1
		0.00%	0.00%	3.31%	3.55%	0.14%	1.60%	0.00%	0.03%	9.83%	0.25%	0.00%	5.00%
Chipotle Peppers	2	0	0	8	85	736	12	2	748	4.4	5.8	0.4	0
		0.00%	0.00%	1.14%	2.58%	32.00%	0.60%	0.73%	14.96%	7.33%	4.83%	2.00%	0.00%
Garlic - Fresh	3	0	0	13.8	36.1	1.5	13	3	0.8	2.8	0.2	0	0.1
		0.00%	0.00%	1.97%	1.09%	0.07%	0.65%	1.09%	0.02%	4.67%	0.17%	0.00%	5.00%
Brown Sugar	1 Tbsp	0.00	0.00	0.00	0.00	0.00	48.38	12.50	0.00	0.00	0.00	0.00	0.00
		0.00%	0.00%	0.00%	0.00%	0.00%	2.42%	4.55%	0.00%	0.00%	0.00%	0.00%	0.00%
Veg. oil	1/2 T	6.81	0.50	0.00	0.00	0.00	60.22	0.00	0.00	0.00	4.84	1.19	0.00
		10.48%	2.27%	0.00%	0.00%	0.00%	3.01%	0.00%	0.00%	0.00%	4.04%	5.95%	0.00%
Lime juice	2 T	0.00	0.00	3.99	32.80	0.60	7.00	2.00	14.00	60.00	0.20	0.10	0.00
		0.00%	0.00%	0.57%	0.99%	0.03%	0.35%	0.73%	0.28%	100.00%	0.17%	0.50%	0.00%
Cumin	1 T	1	0	29.9	107	10.10	22	3	76.2	0.5	0.3	0.2	2
		1.54%	0.00%	4.27%	3.24%	0.44%	1.10%	1.09%	1.52%	0.83%	0.25%	1.00%	100.00%
Salt (Pinch)	1/8 tsp	0				287.50							
						12.50%							
Total for Dish		9.81	0.50	205.19	1647.9	1327.40	392.09	64.00	15674.	91.20	72.34	3.69	2.65
1 1/2 cups		15.10%	2.27%	29.31%	49.94%	57.71%	19.60%	23.27%	313.49	152.00%	60.29%	18.45%	132.50%
# of Servings		16	16	16	16	16	16	16	16	16	16	16	16
2 Tbsp													
Totals per serving		0.61	0.03	12.82	102.99	82.96	24.51	4.00	979.66	5.70	4.52	0.23	0.17
Yield		0.94%	0.14%	1.83%	3.12%	3.61%	1.23%	1.45%	19.59%	9.50%	3.77%	1.15%	8.28%

Essential Amino Acid Chart for Chile Paste

	Mg/ gram protein needed	Pepper Dried Chili	Recipe 6 g	Stock	Recipe 2.35 g	Onion	Recipe .9 g	Chipotle	Recipe .4 g	Total Gram	Total Amino Acid	Tot. p/g
Essential Amino Acids		p/g	Recipe	p/g	Recipe	p/g	Recipe	p/g	Recipe			
Histidine	18	20.70	124.20			12.44	11.20	20.00	8.00	9.65	143.40	14.86
Isoleuciine	25	31.93	191.58			11.79	10.61	33.00	13.20		215.39	22.32
Methionine	25	11.87	71.22			1.68	1.52	13.00	5.20		77.94	8.08
Leucine	55	51.67	310.02			21.05	18.95	53.00	21.20		350.17	36.29
Lysine	51	44.00	264.00			32.84	29.56	46.00	18.40		311.96	32.33
Phenylalanine	47	30.53	183.18			22.22	20.00	32.00	12.80		215.98	22.38
Threonine	27	36.33	217.98			17.68	15.92	36.50	14.60		248.50	25.75
Tryptophan	7	12.60	75.60			12.44	11.20	13.00	5.20		92.00	9.53
Valine	32	41.67	250.02			18.67	16.80	43.00	17.20		284.02	29.43
Amount per recipe		2 oz.		1/2 cup		1/2 cup		2 oz.				
Protein per recipe (g)		6		2.35		0.9		0.4				

Total recipe Protein (g)	# of servings	Protein / serving (g)	Not a Complete Protein
9.65	4	2.41	

Coleslaw

Coleslaw is a mix of julienned or chopped cabbage, along with other crisp vegetables. Carrots are usually a standard that add some sweetness to the tart cabbage. Other options include peppers, fennel, broccoli, apples, etc.

Cabbage is also high in many vitamins and minerals that the body needs and low in potassium and phosphorus, minerals that need to be controlled in the CKD diet. The dressing will have some fat, but as the recipe shows, the benefits outweigh the small amounts of saturated fat.

Health Note

Vitamins are either water soluble or fat soluble. This means that in order to get the best absorption rate they will need to be carried by one or the other. You can stuff yourself with vitamin A, but without some fat, only small amounts will absorb into the body.

Fat Soluble Vitamins: A, D, E, K

Water Soluble Vitamins: B's, C

Ingredients:

- 6 cups green cabbage, shredded, julienned, or chopped
- 1 cup red cabbage, shredded, julienned, or chopped
- 1 cup carrots, grated or peeled into strips
- 1/2 cup coleslaw dressing (see recipe)
- Black Pepper

Total time: 20 minutes Yield: 8 cups Portions: 8
Serving Size: 1 cup

The only instruction you need for this recipe is to combine all the ingredients in a large bowl and mix well.

It will make enough for 8 servings, and it will last for several days to a week. Over time, however, the cabbage will release some of it's water content and you may find that a liquid has formed on the bottom of the storage bowl, mixed with the dressing. You can remix the coleslaw and it will be fine.

Alternatively, you could reduce a small amount of your fluid intake by placing the raw shredded cabbage in a bowl and add a pinch of salt. Let it rest overnight and you will find that some of the water has drained out of the cabbage. Remove the cabbage from the bowl, leaving the water, and mix as directed above.

Health Note

The combination of carrots and cabbage make coleslaw very high in vitamins A,C, and K. Beta carotene is one of the carotenoids found in carrots and the body transforms this to vitamin A. It then takes a fat to assist the absorption into the body, along with the other vitamins. The fat content of the dressing in this recipe aids in the absorption of these ("Vitamin A. Fact sheet for health professionals," n.d.).

Coleslaw

		Fat (g)	Sat. Fat (g)	Phos (mg)	Pot (mg)	Sodium (mg)	Calories	Carbs (g)	Vit A (iu)	Vit C (mg)	Vit K (Mcg)	Vit E (mg) AT	Vit B6 (mg)
	RDA	65	20	700	3300	2300	2000	110	5000	60	120	20	2
Ingredients	Amt												
Green Cabbage	6 Cups	0	0	138.6	906	96	110	6	523.2	195.6	405.6	0.6	0.6
		0.00%	0.00%	19.80%	27.45%	4.17%	5.50%	5.45%	10.46%	326.00%	338.00%	3.00%	30.00%
Red Cabage	1 Cup	0	0	32.10	151	16	33	6	87.2	32.6	67.6	0.1	0.1
		0.00%	0.00%	4.59%	4.58%	0.70%	1.65%	5.45%	1.74%	64.80%	56.33%	0.50%	5.00%
Cole Slaw Dressing	1/2 cup	110.50	10.67	54.00	336.44	256.28	1146.97	9.33	681.48	8.60	78.76	16.13	0.10
		170.00%	53.33%	7.71%	10.20%	11.14%	57.35%	8.48%	13.63%	14.33%	65.64%	80.67%	5.00%
Carrots - Peeled	1 Med	0.00	0.00	38.50	352.00	75.90	52.00	12.00	21363.00	7.60	16.90	0.80	0.20
(61 g or 1 cup)		0.00%	0.00%	5.50%	10.67%	3.30%	2.60%	10.91%	427.26%	12.67%	14.08%	4.00%	10.00%
Total for Dish		110.50	10.67	263.20	1745.44	444.18	1341.97	33.33	22654.88	244.40	568.86	17.63	1.00
		170.00%	53.33%	37.60%	52.89%	19.31%	67.10%	30.30%	453.10%	407.33%	474.05%	88.17%	50.00%
# of Servings		8	8	8	8	8	8	8	8	8	8	8	8
(1 1/2 oz.)													
Totals per serving		13.81	1.33	32.90	218.18	55.52	167.75	4.17	2831.86	30.55	71.11	2.20	0.13
		21.25%	6.67%	4.70%	6.61%	2.41%	8.39%	3.79%	56.64%	50.92%	59.26%	11.02%	6.25%

Essential Amino Acid Chart for Coleslaw

	Mg/ gram protein needed	Cab.	Recipe 6.1 g	Apple Cider Vinegar	Recipe 1.1 g	Cole Slaw Dres.	Recipe 10 g	Carrot	Recipe .8 g	Total Gram	Total Amin o Acid	Total p/g
Essential Amino Acids		**p/g**	**Recipe**	**p/g**	**Recipe**	**p/g**	**Recipe**	**p/g**	**Recipe**			
Histidine	**18**	17.82	117.60	17.82	19.60	7.87	19.99	21.67	17.33	11.04	174.52	15.81
Isoleuciine	**25**	24.27	160.20	24.27	26.70	15.62	39.67	56.67	45.33		271.90	24.63
Methionine	**25**	9.73	64.20	9.73	10.70	6.61	16.78	10.00	8.00		99.68	9.03
Leucine	**55**	33.00	217.80	33.00	36.30	22.12	56.19	58.33	46.67		356.95	32.33
Lysine	**51**	35.64	235.20	35.64	39.20	20.65	52.46	55.00	44.00		370.86	33.59
Phenylalani ne	**47**	25.91	171.00	25.91	28.50	16.73	42.50	43.33	34.67		276.67	25.06
Threonine	**27**	28.36	187.20	28.36	31.20	14.42	36.63	51.67	41.33		296.36	26.84
Tryptophan	**7**	8.91	58.80	8.91	9.80	4.24	10.77	15.00	12.00		91.37	8.28
Valine	**32**	34.00	224.40	34.00	37.40	18.04	45.82	60.00	48.00		355.62	32.21
Recipe Amount		6 Cups		1 Cup		1/2 cup		1 cup				
Protein per recipe (g)		6.6		1.1		2.54		0.8				

Total recipe Protein (g)	# of servings	Protein / serving (g)	Not a Complete Protein
11.04	8	1.38	

Coleslaw Dressing

Making your own dressing for coleslaw is pretty easy, and it allows you to control the sugars and sodium level in your food. Based on my review of some dressings in the refrigerated section of grocery stores, making your own can be much healthier.

In this recipe, for the same reasons as stated above, I have used a homemade aioli recipe. Aioli is similar to mayonnaise (see recipe below). The other reason for homemade dressing, or aioli for that matter, is that it is quite easy to enhance the flavors with "better for you" ingredients. For instance, using fresh onion or shallots is a better health option than powdered or other processed food stuffs.

In the aioli, I use fresh garlic, basil, and parsley to enhance the base flavor of the end product, which can assist in limiting additional salt.

Ingredients:

- 1 cup homemade garlic aioli (or mayonnaise)
- 3 tablespoons sour cream
- 2 tablespoons pickle relish (more if you like)
- 1 tablespoon fresh lemon juice
- 2 teaspoons sugar
- 1 teaspoon ground mustard
- 1 garlic clove, minced
- 1 tablespoon champagne vinegar or white wine vinegar (not distilled white)
- Pinch of salt
- Pepper to taste

Total time: 5 minutes Yield: 1 1/2 cups Portions: 16
Serving size: 1 1/2 Tbsp

Mix all the ingredients together and taste for preference. You may want to add a little more zing (sour cream), a bit more freshness (lemon juice), or sugar to make it sweeter. This is a good base and you should experiment with the ingredients to fine tune the recipe to your taste.

It really is that easy. And let me repeat, by adding more of those "better for you" ingredients, or additional flavors, you will enhance flavor without adding salt.

Food Note

Salt is NaCl or Sodium Chloride. It is 1 part sodium to 1 part chloride. However, salt is made up of about 39% sodium and 61% chloride by mass. This is because the chloride is heavier. Some food item labels will state that there is no sodium in a product, but most often they will use a different chemical form. Usually this is KCl, or Potassium Chloride, and it is 1 part Potassium and 1 part chloride. This can be very dangerous for CKD patients as high potassium levels often need to be avoided.

So read your labels! It may say low or no sodium, but the substitutes can be just as dangerous and often much worse.

Coleslaw Dressing

		Fat (g)	Sat. Fat	Phos (mg)	Pot (mg)	Sodium (mg)	Calorie	Carb (g)	Vit A (iu)	Vit C (mg)	Vit K (Mcg)	Vit E (mg) AT	Vit B6 (mg)
	RDA	**65**	**20**	**700**	**3300**	**2300**	**2000**	**110**	**5000**	**60**	**120**	**20**	**2**
Ingredient	**Amt**												
Aoili, Mayo	**1 cup**	221	32	72.10	492.2	30.80	2040	0.56	297.45	8.4	130.5	31.5	0.1
		340.00	160.00%	10.30%	14.92%	1.34%	102.00%	0.51%	5.95%	14.00%	108.75%	157.50%	5.00%
Pickle Relish	**2 Tbsp**	0	0	4.2	7.4	244	40	0	366	0.4	25.2	0.2	0
		0.00%	0.00%	0.60%	0.22%	10.61%	54.00%	54.00%	7.32%	64.80%	21.00%	1.00%	0.00%
Lemon Juice - fresh	**1 oz**	0.00	0.00	8.00	34.70	1.20	14.00	3.00	14.00	8.40	0.20	0.10	0.00
(2 Tbsp)		0.00%		3.99%	1.05%	0.60%	0.70%	2.73%	0.28%	14.00%	0.17%	0.50%	0.00%
Sugar	**2 Tsp**	0.00	0.00	0.00	0.00	0.00	30.00	25.00	0.00	0.00	0.00	0.00	0.00
		0.00%	0.00%	0.00%	0.00%	0.00%	1.50%	22.73%	0.00%	0.00%	0.00%	0.00%	0.00%
Sour Cream	**3 Tbsp**	16.50		27.75	33.38	16.95	153.94	0.75	656.06	0.00	1.43	0.47	0.00
Made w/ cream		25.38%	0.00%	3.96%	1.01%	0.74%	7.70%	0.68%	13.12%	0.00%	1.19%	2.34%	0.00%
Ground Mustard	**1 tsp**	0.00	0.00	0.00	0.00	0.00	0.00	0.00	0.00	0.00	0.00	0.00	0.00
		0.00%	0.00%	0.00%	0.00%	0.00%	0.00%	0.00%	0.00%	0.00%	0.00%	0.00%	0.00%
Garlic clove	**1 med**	0		4.6	12	0.5	13	1	29.45	0.9	0.2	0	0.1
		0.00%	0.00%	0.66%	0.36%	0.02%	0.65%	0.91%	0.59%	1.50%	0.17%	0.00%	5.00%
Vinegar	**1 Tbsp**	0.00		19.10	93.20	19.10	3.00	0.13	0.00	0.00	0.00	0.00	0.00
		0.00%	0.00%	2.73%	2.82%	0.83%	0.15%	0.11%	0.00%	0.00%	0.00%	0.00%	0.00%
Salt	**Pinch**	0.00		0.00	0.00	200.00							
		0.00%	0.00%	0.00%	0.00%	8.70%	0.00%	0.00%	0.00%	0.00%	0.00%	0.00%	0.00%
Total for Dish		237.50	32.00	135.75	672.88	512.55	2293.94	30.44	1362.96	18.10	157.53	32.27	0.20
		365.38	160.00%	19.39%	20.39%	22.28%	114.70%	27.67%	27.26%	30.17%	131.27%	161.34%	10.00%
# of Servings		16	16	16	16	16	16	16	16	16	16	16	16
Totals per serving		14.84	2.00	8.48	42.05	32.03	143.37	1.90	85.19	1.13	9.85	2.02	0.01
Makes 1 1/2 cups		22.84%	10.00%	1.21%	1.27%	1.39%	7.17%	1.73%	1.70%	1.89%	8.20%	10.08%	0.63%

Essential Amino Acid Chart for Coleslaw dressing

	Mg/ gram protein needed	Aioli/ Mayo	Recipe 3.8 g	Lemon Juice	Recipe .1 g	Sour Cream	Recipe .91 g	Fresh Garlic	Recipe .2 g	Pickle Relish	Recipe .8 g	Total Gram	Total Amino Acid	Total P/g
Essential Amino Acids		p/g	Recipe	p/g	Recipe	p/g	Recipe	p/g	Recipe	p/g	Recipe			
Histidine	18	25.59	97.23	38.70	3.87	27.14	24.70	17.00	3.40	21.67	17.33	5.81	121.84	20.97
Isoleucine	25	53.72	204.12	53.00	5.30	60.20	54.78	32.50	6.50	56.67	45.33		261.25	44.97
Methionine	25	20.33	77.27	26.80	2.68	24.89	22.65	11.50	2.30	10.00	8.00		90.25	15.53
Leucine	55	79.93	303.74	96.70	9.67	97.55	88.77	47.00	9.40	58.33	46.67		369.48	63.59
Lysine	51	70.70	268.67	92.58	9.26	78.78	71.69	42.00	8.40	55.00	44.00		330.33	56.85
Phenylalanine	47	42.51	161.54	53.80	5.38	48.16	43.83	27.50	5.50	43.33	34.67		207.09	35.64
Threonine	27	43.87	166.71	36.90	3.69	45.10	41.04	23.50	4.70	51.67	41.33		216.44	37.25
Tryptophan	7	11.96	45.46	13.50	1.35	14.08	12.81	10.00	2.00	15.00	12.00		60.81	10.47
Valine	32	59.10	224.57	68.70	6.87	66.53	60.54	43.50	8.70	60.00	48.00		288.14	49.59
Recipe Amount		1 Cup		1 each		2 Tbsp		1 clove		1 cup				
Protein per recipe (g)		3.8		0.1		0.91		0.2		0.8				

Total recipe Protein (g)	# of servings	Protein / serving (g)	Not a Complete Protein
5.81	16.00	0.36	

Homemade Garlic Aioli

Why? Why would someone want to make their own mayonnaise? The truth is not many people do. However, it is a base ingredient in many dishes, and having more control over the key ingredients in any recipe has benefits for taste, and for better health.

What about the Fat? Yes, the fat. Lots of it, but we do need some fat in our diet, and if we can bump up the flavors of this ingredient, we will use less of it to enjoy our final dishes. Store bought mayonnaise is intentionally neutral in flavor, so you end up adding fat without much flavor. Not a good trade off.

We need a few definitions first. Mayonnaise is an emulsion made of some kind of oil, egg yolk, lemon, vinegar, and mustard. Aioli could be described as a type of mayonnaise, but it must be made with olive oil and garlic, preferably fresh minced garlic.

Most mayonnaise products are made with neutral oils like canola or soy. The mild lemony and vinegar notes may come through, but it is mostly neutral.

Aioli is meant to be flavorful in itself. Olive oil alone adds flavor unlike the neutral oils, and of course we all know about fresh garlic. The addition of other aromatics serve to enhance the end product even further.

Food Note

Mayonnaise and Aioli are types of emulsions.This a suspension of liquids that do not mix together naturally. Some emulsions will only last a short time without a chemical stabilizer - like a vinaigrette - while others will hold together indefinitely - like a mayonnaise. The process is to break of up one of the liquids into smaller droplets using a force (whisking), and combining an emulsifier (lecithin in the egg yolk and mustard) to get in between the droplets so that the liquids will not return to their original state. One end of the lecithin binds to fat and the other to water. This is the magic that holds these two ingredients together indefinitely.

In the diet program for CKD patients, it is recommended that recipes be enhanced with flavors and ingredients that are healthier so that unhealthy items like salt and fat are used less.

Ingredients:

- 1 cup olive oil
- 1 egg yolk from a pasteurized egg*
- 1 ounce fresh lemon juice
- 1 teaspoon ground mustard
- 1 medium garlic clove, minced
- 1 tablespoon white or champagne vinegar - not that distilled white stuff
- Pinch of salt
- Black pepper to taste

Time: 10 minutes Yield: 1 Cup Portions: 16
Serving size: 1 Tbsp

Start by finding a non-reactive bowl. Glass, stainless steel, ceramic are all good choices. Cast iron, aluminum, and copper are reactive bowls and should be avoided. Reactive bowls can pick up some metallic flavorings, especially with acidic foods or when used to cook something over a long period of time.

Food Note

Creating a suspension requires breaking down of one of the liquids - a fat in this case - very slowly. For this process, starting with a few drops of oil is much easier than trying to take on large amounts. Too much at once and it will not happen at all. Once the lecithin starts to connect to the droplets of water and fat, gradually add more oil. As you notice your mixture start to look like mayonnaise, you can add more of the oil at one time.

Place the egg yolk in the bowl and whisk well. Next add half of the lemon juice, vinegar and all of the ground mustard (it also has lecithin needed to bind). Mix this well. Now it's time to start adding the oil, very slowly at first. You will need to whisk constantly and vigorously in the beginning until a suspension has started. After a while you can slow down the mixing, but keep it steady.

Start with a few drops, whisk, and then add another few drops. Mix again. When the mixture starts to look like it is binding you can add more oil, a little at a time. If you add too much, you could break the whole thing down and will have to start over.

About half way through the addition of the oil, add the rest of the ingredients and continue to whisk. When finished you will have a fragrant, smooth, silky and delicious aioli.

Health Note

Eggs have long been called the perfect food. This reference has to do with the amino acids that make up the protein, the availability, and the cost.

Eggs provide all the essential amino acids at levels to be considered a complete protein. The essential amino acids are those that the body cannot produce itself and must be consumed from outside sources. Inexpensive and available, eggs were often eaten to maintain health when meat wasn't readily available. Many immigrants (my grandfather included) used to put a raw egg in their morning coffee, and call it breakfast. It is also why Rocky guzzles all those raw eggs in the morning. An old style and less expensive version of today's protein powder, albeit a bit risky.

*I am not recommending that you eat raw eggs. This was only meant as an anecdote.

*If you cannot find pasteurized eggs at the store, you can make your own. Place eggs in water, bring temperature up to 140 ° and maintain for 5 minutes. This will kill the potential bacteria.

Homemade Aioli (Mayonnaise)

		Fat (g)	Sat. Fat	Phos (mg)	Pot (mg)	Sodium (mg)	Calories	Carb (g)	Vit A (iu)	Vit C (mg)	Vit K (Mcg)	Vit E (mg) AT	Vit B6 (mg)
	RDA	65	20	700	3300	2300	2000	110	5000	60	120	20	2
Ingredients	Amt												
Olive oil	1 cup	216	30	2.2	4.3	0	1910	0	0	0	130	31	0
		332.31	150.00%	0.31%	0.13%	0.00%	95.50%	0.00%	0.00%	0.00%	108.33%	155.00%	0.00%
Egg Yolk	1 each	5	2	38.2	348	8	100	1	254	0	0.1	0.4	0
1 large (17 g)		7.69%	10.00%	5.46%	10.55%	0.35%	54.00%	0.91%	5.08%	64.80%	0.08%	2.00%	0.00%
Lemon Juice - fresh	1 oz	0.00	0.00	8.00	34.70	1.20	14.00	3.00	14.00	8.40	0.20	0.10	0.00
(2 Tbsp)		0.00%	0.00%	3.99%	1.05%	0.60%	0.70%	2.73%	0.28%	14.00%	0.17%	0.50%	0.00%
Ground Mustard	1 tsp	0.00	0.00	0.00	0.00	0.00	0.00	4.00	0.00	0.00	0.00	0.00	0.00
		0.00%	0.00%	0.00%	0.00%	0.00%	0.00%	3.64%	0.00%	0.00%	0.00%	0.00%	0.00%
Garlic clove	1 med	0	0	4.6	12	0.5	13	1	29.45	0.9	0.2	0	0.1
		0.00%	0.00%	0.66%	0.36%	0.02%	0.65%	0.91%	0.59%	1.50%	0.17%	0.00%	5.00%
Vinegar - Champagne	1 Tbsp	0.00	0.00	19.10	93.20	19.10	3.00	0.00	0.00	0.00	0.00	0.00	0.00
		0.00%	0.00%	2.73%	2.82%	0.83%	0.15%	0.00%	0.00%	0.00%	0.00%	0.00%	0.00%
Salt	1/2 tsp	0.00	0.00	0.00	0.00	1150.00	0.00	0.00	0.00	0.00	0.00	0.00	0.00
		0.00%	0.00%	0.00%	0.00%	50.00%	0.00%	0.00%	0.00%	0.00%	0.00%	0.00%	0.00%
Total for Dish		221.00	32.00	72.10	492.20	1178.80	2040.00	9.00	297.45	9.30	130.50	31.50	0.10
		340.00	160.00%	10.30%	14.92%	51.25%	102.00%	8.18%	5.95%	15.50%	108.75%	157.50%	5.00%
# of Servings		16	16	16	16	16	16	16	16	16	16	16	16
Totals per serving		13.81	2.00	4.51	30.76	73.68	127.50	0.56	18.59	0.58	8.16	1.97	0.01
		21.25%	10.00%	0.64%	0.93%	3.20%	6.38%	0.51%	0.37%	0.97%	6.80%	9.84%	0.31%

Essential Amino Acid Chart for Aioli

	Mg/ gram protein needed	Egg Yolk	Recipe 2.7 g	Lemon Juice	Recipe .1 g	Must.	Recipe .9 g	Fresh Garlic	Recipe .2 g	Carrot	Recipe .8 g	Total Gram	Total Amino Acid	Total P/g
Essential Amino Acids		p/g	Recipe	p/g	Recipe	p/g	Recipe	p/g	Recipe	p/g	Recipe			
Histidine	18	26.90	72.63	38.70	3.87	31.04	27.93	17.00	3.40	21.67	17.33	4.7	97.23	20.69
Isoleuciine	25	54.44	146.99	53.00	5.30	44.07	39.67	32.50	6.50	56.67	45.33		204.12	43.43
Methionine	25	23.81	64.29	26.80	2.68	19.56	17.60	11.50	2.30	10.00	8.00		77.27	16.44
Leucine	55	88.15	238.01	96.70	9.67	72.59	65.33	47.00	9.40	58.33	46.67		303.74	64.63
Lysine	51	76.67	207.01	92.58	9.26	61.85	55.67	42.00	8.40	55.00	44.00		268.67	57.16
Phenylalanine	47	42.96	115.99	53.80	5.38	43.33	39.00	27.50	5.50	43.33	34.67		161.54	34.37
Threonine	27	43.33	116.99	36.90	3.69	44.44	40.00	23.50	4.70	51.67	41.33		166.71	35.47
Tryptophan	7	11.15	30.11	13.50	1.35	21.44	19.30	10.00	2.00	15.00	12.00		45.46	9.67
Valine	32	59.63	161.00	68.70	6.87	54.07	48.67	43.50	8.70	60.00	48.00		224.57	47.78
Recipe Amount		1		1 each		1 tsp		1 clove		1 cup				
Protein per recipe (g)		2.7		0.1		0.9		0.2		0.8				

Total recipe Protein (g)	# of servings	Protein / serving (g)	Not a Complete Protein
4.70	16.00	0.29	

Barbecue Sauce

Summer time means grilling, BBQ, and picnics. CKD patients can have a difficult time navigating all the wonders of the outdoor grilling season. These include dry rubs with high sodium levels; wet sauces that are high in potassium, sugars, and sodium; and high fat meats that have excessive phosphorus levels. All of these are on the "avoid or limit" list.

This recipe accommodates many of the problem areas and can relieve some of the concerns that occur when trying to enjoy the outdoor cooking season. The use of tomato paste helps with the potassium levels; limited brown sugar reduces carbs and sugar levels; minimal salt limits sodium; and using the chili paste listed above, greatly enhances the flavor without adding to the negative aspects of jarred BBQ sauce.

If you choose to use a pre-made chili paste, I recommend adding 2 chipotle peppers for the smokey flavor, and find one with the least amount of sodium and sugar.

Ingredients:

- 1/2 cup chili paste
- 1/4 cup apple cider vinegar
- 2 tablespoon tomato paste (no salt if available)
- 1/3 cup brown sugar
- 2 tablespoons vegetable oil (canola or other neutral oil)
- 1 clove garlic, minced
- 1/2 cup onion, chopped
- 1/2 tablespoon ground mustard
- 1/2 teaspoon salt
- Black pepper to taste

Total time: 40 min. Yield: About 1 1/2 cups Portions: 12
Serving size: 2 Tbsp.

Warm a sauté pan large enough to hold all the ingredients. When the pan is ready, add the oil and wait a few seconds for the oil to start to shimmer. This means that the oil is hot and ready. Add the chopped onion and let sweat for about 3 minutes, or until the onions are translucent. Next add the garlic and cook for less than a minute.

Add the tomato paste and mix with the onion/garlic mixture. Cook for 1 to 2 minutes. Cooking the paste in the onion/garlic will deepen the flavor even more.

Add the rest of the ingredients and mix thoroughly, leaving the sauce on medium heat for another 5 minutes.

Here you can adjust the flavor for your particular taste; spicier (more chili paste or cayenne pepper), sweeter (more brown sugar), smokier (add chipotle pepper). Just remember that these add-ons will alter the nutritional data provided for the recipe.

BBQ Sauce (Homemade)

		Fat (g)	Sat. Fat (mg)	Phos (mg)	Pot (mg)	Sodium (mg)	Calorie	Carb (g)	Vit A (iu)	Vit C (mg)	Vit K (Mcg)	Vit E (mg) AT	Vit B6 (mg)
	RDA	65	20	700	3300	2300	2000	110	5000	60	120	20	2
Ingredients	Amt												
Chile Paste	1/2 Cup	10.08	0.67	68.40	549.30	442.47	190.92	0.53	5224.87	30.40	28.96	2.42	0.88
		15.51%	3.33%	9.77%	16.65%	19.24%	9.55%	0.48%	104.50%	50.67%	24.13%	12.10%	44.17%
Apple Cider Vin.	1/4 cup	0	0	4.775	43.50	3	12.5	0.5	0.00	0.00	0.00	0.00	0.00
		0.00%	0.00%	0.68%	1.32%	0.13%	0.63%	0.45%	0.00%	64.80%	0.00%	0.00%	0.00%
Tomato Paste	2 Tbsp	0.00	0.00	23.20	284.00	221.00	23.00	5.00	427.00	6.10	3.20	1.20	0.10
(1 oz.)		0.00%	0.00%	3.31%	8.61%	9.61%	1.15%	4.55%	8.54%	10.17%	2.67%	6.00%	5.00%
Brown Sugar	1/3 Cup	0.00	0.00	2.93	97.67	20.53	287.67	72.00	0.00	0.00	0.00	0.00	0.00
		0.00%	0.00%	0.42%	2.96%	0.89%	14.38%	65.45%	0.00%	0.00%	0.00%	0.00%	0.00%
Garlic clove	1 cove	0.00	0.00	4.60	12.03	0.50	4.30	1.00	0.27	0.93	0.07	0.00	0.03
		0.00	0.00	0.01	0.00	0.00	0.00	0.01	0.00	0.02	0.00	0.00	0.02
Vegetable oil	2 Tbsp	27.00	2.00	0.00	0.00	0.00	240.88	0.00	0.00	0.00	19.38	4.76	0.00
		0.00%	0.00%	0.66%	0.36%	0.02%	0.22%	0.91%	0.01%	1.56%	0.06%	0.00%	1.67%
Chopped onion	1/2 cup	0.00	0.00	23.20	117.00	3.20	32.00	0.00	1.60	59.00	0.30	0.00	0.10
		0.00%	0.00%	3.31%	3.55%	0.14%	1.60%	0.00%	0.03%	98.33%	0.25%	0.00%	5.00%
Ground Must.	1/2 Tbsp	0.02	0.00	46.25	32.50	0.30	26.00	2.00	3.40	0.15	0.30	0.15	0.00
		0.02%	0.00%	6.61%	0.98%	0.01%	1.30%	1.82%	0.07%	0.25%	0.25%	0.75%	0.00%
Salt	1 /2 Tsp	0.00				1150.00							
		0.00%	0.00%	0.00%	0.00%	50.00%	0.00%	0.00%	0.00%	0.00%	0.00%	0.00%	0.00%
Total for Dish		37.10	2.67	173.36	1136.00	1841.00	817.26	81.03	5657.13	96.58	52.20	8.53	1.12
1 1/2 cups		57.07%	13.33%	24.77%	34.42%	80.04%	40.86%	73.67%	113.14%	160.97%	43.50%	42.67%	55.83%
# of Servings		12	12	12	12	12	12	12	12	12	12	12	12
2 Tbsp/serving													
Totals per serving		3.09	0.22	14.45	94.67	153.42	68.10	6.75	471.43	8.05	4.35	0.71	0.09
		4.76%	1.11%	2.06%	2.87%	6.67%	3.41%	6.14%	9.43%	13.41%	3.63%	3.56%	4.65%

Essential Amino Acid Chart for BBQ Sauce

	Mg/ gram protein needed	Chile Paste	Recipe .82 g	Fresh Onion	Recipe .9 g	Fresh Garlic	Recipe 2.7 g	Tom. Paste	Recipe .8 g	Total Grams	Total	Total Protein/g
Essential Amino Acids		p/g	Recipe	p/g	Recipe	p/g	Recipe	p/g	Recipe			
Histidine	18	15.12	12.40	12.44	11.20	17.00	5.10	13.50	11.07	2.84	39.77	14.00
Isoleuciine	25	23.11	18.95	11.79	10.61	32.50	9.75	16.58	13.60		52.91	18.63
Methionine	25	8.42	6.91	1.68	1.52	11.50	3.45	4.17	3.42		15.29	5.38
Leucine	55	37.56	30.80	21.05	18.95	47.00	14.10	26.17	21.46		85.30	30.04
Lysine	51	33.30	27.30	32.84	29.56	42.00	12.60	28.67	23.51		92.97	32.73
Phenylalanine	47	22.92	18.80	22.22	20.00	27.50	8.25	28.25	23.17		70.21	24.72
Threonine	27	26.15	21.44	17.68	15.92	23.50	7.05	24.00	19.68		64.09	22.57
Tryptophan	7	9.63	7.90	12.44	11.20	10.00	3.00	5.83	4.78		26.88	9.46
Valine	32	30.21	24.77	18.67	16.80	43.50	13.05	17.50	14.35		68.97	24.29
Recipe Amount		2 Tbsp		1/2 cup		1 clove		2 Tbsp				
Protein per recipe (g)		0.82		0.9		0.3		0.82				

Total recipe Protein (g)	# of servings	Protein / serving (g)	Not a Complete Protein
2.84	12.00	0.24	

Chicken French

Chicken French is a chicken cutlet finished in a butter, wine, and lemon sauce. A chicken cutlet is breast meat that has been pounded thin or in the case of this recipe, split in half lengthwise, lightly coated and pan fried. It is a derivation of Veal French, though not french in origin. The French in the name refers to the butter and wine sauce.

Not only is it delicious, but it is high in protein. It has about 33% of the RDI of phosphorus for a 3 oz portion, but the phosphorus to protein ratio is 7.3.

Health Note

Protein and phosphorus intake are related. The more of one, the higher the other. It is often recommended that protein intake be reduced for CKD stages 1 through 4, and then increased at stage 5. At this point patients often need to increase protein and limit phosphorus. The ideal protein will have a low phosphorus to protein ratio, that is high protein levels and low phosphorus. Professionals now discuss the ratio to evaluate food items and recipes that are the most beneficial, especially for dialysis patients.

Ingredients:

- 2 pounds boneless skinless chicken breast, split horizontally so they are even and thin.
- 1 cup all purpose flour (1/2 cup used for nutrition data, because of waste)
- 3 eggs
- 1/4 cup freshly grated Parmigiano Reggiano
- 1/2 cup olive oil or vegetable oil for frying
- 4 tablespoons unsalted butter (1/2 stick)
- 2 lemons, zested and juiced
- 1/2 cup Sherry wine, or other dry white wine
- 2 cloves fresh garlic (optional, I prefer mine without)
- 1/2 teaspoon Salt
- 1 lemon for garish (optional)

Time: 45 minutes   Yield: 2 lbs.   Portions: 8   Serving size: 4oz.

This is a large recipe because the final product holds up well in the fridge for several days. One of the problems I encountered while on dialysis was the motivation and energy to cook on treatment days. Having these leftovers helped me eat well when I was down and out after a day of dialysis.

First on the list is to prepare your breading station. In this recipe there are only two parts, but we will dip three times. Place your eggs in one bowl, and whisk. Combine your flour, salt, pepper, and Parmesan in the next bowl and mix thoroughly.

Set a frying pan, large enough to hold half of the cut chicken, on medium high heat. While the pan is warming, you can prep your chicken breasts. Slice them in half horizontally making them as consistent as possible for even cooking. Alternatively, you could pound out the chicken between two pieces of plastic wrap to get the same results.

Dip the chicken in the flour mixture and shake off any excess flour. Next, dip in the eggs, let drain, and then back in the flour. Place the chicken on a separate plate and continue with the chicken until you have enough to fill the frying pan with one layer. Don't crowd the pan too much as it will lower the temperature, and the oil will soak into the chicken.

Food Note

Breading stations, for either frying or baking, have a specific order. Wet things stick to dry things which then stick to wet things. Remember that meat and vegetables are made up of a great deal of water. The first dip is usually a flour mixture that sticks to the product, then into the egg mixture to stick to the flour, and finally to another dry ingredient like bread crumbs, or in this case flour again.

Cook for about 5-6 minutes per side, or until the chicken reaches a golden brown color. I like to prepare the next batch of chicken while the first is frying. Flip the chicken only once, and remove to a clean plate when done. At this point, the chicken may not be cooked all the way through, but is finished cooking in the sauce. It can be prepared ahead of time and finished when you are ready to serve dinner.

The sauce is where we find some controversy for chefs with the recipe. Amounts and type of alcohol vary, and I am a big fan of finding your own sweet spot through experimentation. My recipe is a little lemon friendly, but that's the way I like it.

Using the same frying pan (with most of the excess oil removed, but not the "bits" from the breading) on medium heat, add the butter and let it melt in the pan. When the butter is almost melted, add the lemon zest and juice. Let sit for about 2 minutes and then add the wine, scraping any pieces from the bottom of the pan. Turn up the heat to medium high, and let the sauce reduce and thicken.

The final step is to add the chicken to the sauce and finish cooking the meat through while in the pan, about 5 minutes. Some of the sauce will incorporate into the breading, and the rest should be served over the chicken.

Health Note

Protein is necessary in all our diets. The amino acids that make up protein are the building blocks of life. It helps build and repair cells in our body and provides us with important enzymes and hormones. When our kidneys are not functioning properly, protein intake and the particular essential amino acids need to be monitored and controlled. When the body processes protein it makes waste, and normal kidneys can remove this without much difficulty. In stages 1-4 of CKD, the kidneys cannot remove the waste adequately and this can further damage the kidneys. Reducing protein intake at this stage is often recommended.

When a person reaches stage 5 (ESRD), they are put on dialysis because the kidneys do not function well enough to maintain life. Dialysis removes the waste created by proteins but it also removes amino acids. At this stage it is usually recommended that patients increase their protein intake.

Chicken French

		Fat (g)	Sat. Fat (g)	Phos (mg)	Pot (mg)	Sodium (mg)	Calories	Carbs (g)	Vit A (iu)	Vit C (mg)	Vit K (Mcg)	Vit E (mg) AT	Vit B6 (mg)
	RDA	65	20	700	3300	2300	2000	110	5000	60	120	20	60
Ingredients	Amt												
Eggs	3	15	6	191	134	420	213	0	487	0	0.3	1	0.1
		23.08%	30.00%	27.29%	4.06%	18.26%	10.65%	0.00%	9.74%	0.00%	0.25%	5.00%	0.17%
Lemons 2	4 Tbsp	0	0	6.27	115.434	1	23.166	6	18.612	131.01	1.1	0	0.4
		0.00%	0.00%	0.90%	3.50%	0.04%	1.16%	5.45%	0.37%	64.80%	0.92%	0.00%	0.67%
Garlic, fresh	1 clove	0	0	4.6	12	0.5	4.3	1	0.25	0.9	0.1	0	0
		0.00%	0.00%	0.66%	0.36%	0.02%	0.22%	0.91%	0.01%	1.50%	0.08%	0.00%	0.00%
Butter	4 Tbsp	44	28	13.6	13.6	8	400	0	1400	0	4	1.2	0
		67.69%	140.00%	1.94%	0.41%	0.35%	20.00%	0.00%	28.00%	0.00%	3.33%	6.00%	0.00%
Chicken Breast	2 lbs.	9.6	3.2	1756.8	2284.8	582.4	984	0	188.8	9.6	3.2	0	0.8
		14.77%	16.00%	250.97%	69.24%	25.32%	49.20%	0.00%	3.78%	16.00%	2.67%	0.00%	1.33%
Parm. Regg.	1/4 cup (1 oz)	3	5	194	25.8	449	40	2	112	0	0.5	0.1	0
		4.62%	25.00%	27.71%	0.78%	19.52%	2.00%	1.82%	2.24%	0.00%	0.42%	0.50%	0.00%
Olive Oil	1/4 cup	54	7.5	0.6	1.07	0	477.5	0	0	0	32.5	8	0
(2 g)		83.08%	37.50%	0.09%	0.03%	0.00%	23.88%	0.00%	0.00%	0.00%	27.08%	40.00%	0.00%
Flour, White AP Enriched.	1 cup	1	0.2	135	134	2	455	47.5	0	0	0.4	0.3	0.1
(6 oz. or 168 g)		1.54%	1.00%	19.29%	4.06%	0.09%	22.75%	43.18%	0.00%	0.00%	0.33%	1.50%	0.17%
Sherry Wine	1/2 cup	0.00	0.00	16.80	98.40	700.00	8.00	8.00	0.00	0.00	0.00	0.00	0.00
Salt	1 tsp					1150.00							
						50.00%							
Total for Dish		126.60	49.90	2318.67	2819.10	3312.90	2604.97	64.50	2206.6	141.51	42.10	10.60	1.40
		194.77%	249.50%	331.24%	85.43%	144.04%	130.25%	58.64%	44.13%	235.85%	35.08%	53.00%	2.33%
# of Servings	4 oz. ser.	8	8	8	8	8	8	8	8	8	8	8	8
	3 oz. ser.	10	10	10	10	10	10	10	10	10	10	10	10
Totals per serving	4 0z	15.83	6.24	289.83	352.39	414.11	325.62	8.06	275.83	17.69	5.26	1.33	0.18
		24.35%	31.19%	41.40%	10.68%	18.00%	16.28%	7.33%	5.52%	29.48%	4.39%	6.63%	0.29%
	3 0z	12.66	4.99	231.87	281.91	331.29	260.50	6.45	220.67	14.15	4.21	1.06	0.14
		19.48%	24.95%	33.12%	8.54%	14.40%	13.02%	5.86%	4.41%	23.59%	3.51%	5.30%	0.23%

Essential Amino Acid Chart for Chicken French

	Mg/g Prot.	Lem.	Rec. .3 g	Egg	Rec. 18.9 g	Flour	Rec. 12.9 g	Butter	Rec. .4g	Parm Regg	Rec. 10 g	Chick.	Rec. 203.4	Total Gram	Total Amin o Acid	Total P/g
Essential Amino Acids		p/g	Rec.	p/g	Rec.	p/g	Rec.	p/g	Rec.	p/g	Rec.	p/g	Rec.			
Histidine	**18**	16.90	5.07	24.60	465.00	22.33	144.00	27.47	10.99	38.70	387.00	31.04	6313.54	239.15	7325.60	30.63
Isoleuciine	**25**	36.86	11.06	53.33	1008.00	34.57	223.00	61.05	24.42	53.00	530.00	52.74	10727.3		12523.7	52.37
Methionine	**25**	11.48	3.44	30.16	570.00	17.75	114.50	25.11	10.04	26.80	268.00	27.68	5630.11		6596.10	27.58
Leucine	**55**	44.29	13.29	86.35	1632.00	68.84	444.00	98.95	39.58	96.70	967.00	75.00	15255.0		18350.8	76.73
Lysine	**51**	47.62	14.29	72.54	1371.00	22.09	142.50	80.00	32.00	92.60	926.00	84.94	17276.8		19762.5	82.64
Phenylalanin e	**47**	28.38	8.51	54.13	1023.00	50.39	325.00	49.00	19.60	53.80	538.00	39.63	8060.74		9974.86	41.71
Threonine	**27**	30.76	9.23	44.13	834.00	27.21	175.50	45.42	18.17	36.90	369.00	42.20	8583.48		9989.38	41.77
Tryptophan	**7**	9.81	2.94	13.25	250.50	12.33	79.50	14.32	5.73	13.50	135.00	11.71	2381.81		2855.48	11.94
Valine	**32**	43.29	12.99	68.25	1290.00	40.23	259.50	67.89	27.16	68.70	687.00	49.57	10082.5		12359.1	51.68
Amount per recipe		3 each		3		1/2 cup used		4 Tbsp		1 ounce		2 lbs.				
Protein per recipe (g)		0.3		18.9		6.45		0.4		10		203.4				

Total recipe Protein (g)	**# of servings**	**Protein / serving (g)**	Complete Protein
239.45	8.00	29.93	

Chicken Tenders - Baked

Maintaining a healthy diet can be difficult and exhausting. When the diet is not just about weight loss or feeling better, but about a serious health issue, it can become daunting and overwhelming. It is important to provide options that resemble the eating habits prior to becoming ill.

I spent more than 5 years on dialysis and that experience taught me that patients (myself included) are always going to eat things that we know are bad for us. I saw bags of chips, fast food containers, and sodas in the hands of patients on a regular basis. This made me want to provide an alternative to those choices.

The recipe below removes some of the problematic items found in the frozen section or at restaurants. In addition, by using Panko breadcrumbs, much of the salt and preservatives found in other breadcrumbs is avoided.

Food Note

Panko is a special type of Japanese breadcrumb made from crustless bread. The pieces are larger than standard breadcrumbs and give food a nice crunchy texture. They are made by using electrical current rather than direct heat from a oven, and no browning occurs on the outside of the bread. You may use plain bread crumbs if panko is not available but the end result will not have the crunchy exterior.

Ingredients:

- 2 pounds. raw chicken breast or chicken strips
- 1 1/2 cups panko breadcrumbs (or plain regular)
- 2 large eggs
- 1/4 cup Parmigiano Reggiano
- 3 tablespoons Olive Oil
- 1 cup all purpose flour
- 1/4 teaspoon salt
- 1/4 teaspoon pepper

Time : 45 minutes Yield: 2 pounds Portions: 8
Serving size: 1/4 pound.

This is a standard three-part breading process. Part one is the flour and spices - mix the flour, salt, and pepper in one bowl. Part two - whisk the eggs in a separate bowl. Part three - mix the panko and the Parmigiano Reggiano in the third.

Heat your oven to 425° F. Set up a cooling rack on a sheet pan large enough to hold all the chicken strips when they are coated. Cut the chicken breasts into equal strips if you bought whole breasts. Your stations are now ready.

Dip the chicken in the flour, shake off the excess and place in the eggs. Let the egg drip off the chicken, and then place in the panko mixture, toss until fully coated. Place the fully coated chicken on the rack and repeat with the rest of the chicken.

When all the chicken is breaded, place it in the oven and bake for 25 to 30 minutes, or until the inside reaches 165° F, and the outside is golden brown. When cooked you can put the pan directly in the freezer on the cooling rack. After a few hours the chicken will freeze and you can move it to a plastic bag and keep for up to 3 months.

This is something I like to have in the freezer for those days when I'm hungry but the energy level to cook just is not there. Letting these thaw for 1 hour and then baking for another 30 minutes is easy enough to cook even on a difficult dialysis day.

Food/Health Note

Many of the recipes provided have a large number of servings. I have done this intentionally both because of the effects of dialysis, and the general time constraints of daily life for everyone. My time conducting in-center treatment gave me the opportunity to talk to many different people in the same position. An overwhelming commonality was the exhaustion after dialysis.
It was not just me that found it difficult to move from the couch to the kitchen on those days. When I had the energy to cook, I made sure to prepare a few extra meals to accommodate the lack of energy and the inability to make a decision after treatment.

Chicken Tenders (Baked)

		Fat (g)	Sat. Fat (g)	Phos (mg)	Pot (mg)	Sodium (mg)	Calories	Carbs (g)	Vit A (iu)	Vit C (mg)	Vit K (Mcg)	Vit E (mg) AT	Vit B6 (mg)
		65	20	700	3300	2300	2000	110	5000	60	120	20	60
Ingredients	Amt												
Eggs	2	10.00	4.00	63.67	44.67	143.33	71.00	0.00	162.33	0.00	0.10	0.33	0.03
		15.38%	20.00%	9.10%	1.35%	6.23%	3.55%	0.00%	3.25%	0.00%	0.08%	1.65%	0.05%
Panko Bread Crumbs	1 1/2 cups	8.55	1.95	267	318	141	156	33	18.612	131.01	7.1	0.1	0.1
		13.15%	9.75%	38.14%	9.64%	6.13%	7.80%	30.00%	0.37%	64.80%	5.92%	0.50%	0.17%
Chicken Breast	2 lbs.	9.6	3.2	1756.8	2284.8	582.4	984	0	188.8	9.6	3.2	0	0.8
		14.77%	16.00%	250.97%	69.24%	25.32%	49.20%	0.00%	3.78%	16.00%	2.67%	0.00%	1.33%
Parmigiano Reggiano Fresh	1/4 cup (1 oz)	3	5	194	25.8	449	40	2	112	0	0.5	0.1	0
(2 g)		4.62%	25.00%	27.71%	0.78%	19.52%	2.00%	1.82%	2.24%	0.00%	0.42%	0.50%	0.00%
Olive Oil	3 Tbsp	40.5	5.625	0.4125	0.80625	0	358.125	0	0	0	24.375	5.8125	0
(2 g)		62.31%	28.13%	0.06%	0.02%	0.00%	17.91%	0.00%	0.00%	0.00%	20.31%	29.06%	0.00%
Flour, White AP Enriched.	1 cup	1	0.2	135	134	2	455	47.5	0	0	0.4	0.3	0.1
(6 oz. or 168 g)		1.54%	1.00%	19.29%	4.06%	0.09%	22.75%	43.18%	0.00%	0.00%	0.33%	1.50%	0.17%
Total for Dish		72.65	19.98	2416.88	2808.07	1317.73	2064.13	82.50	481.75	140.61	35.68	6.64	1.03
		111.77%	99.88%	345.27%	85.09%	57.29%	103.21%	75.00%	9.63%	234.35%	29.73%	33.21%	1.72%
# of Servings	4 oz. ser.	8	8	8	8	8	8	8	8	8	8	8	8
	3 oz. ser.	10	10	10	10	10	10	10	10	10	10	10	10
Totals per serving	4 0z	9.08	2.50	302.11	351.01	164.72	258.02	10.31	60.22	17.58	4.46	0.83	0.13
		13.97%	12.48%	43.16%	10.64%	7.16%	12.90%	9.38%	1.20%	29.29%	3.72%	4.15%	0.21%
	30z	7.27	2.00	241.69	280.81	131.77	206.41	8.25	48.17	14.06	3.57	0.66	0.10
		11.18%	9.99%	34.53%	8.51%	5.73%	10.32%	7.50%	0.96%	23.44%	2.97%	3.32%	0.17%

Essential Amino Acid Chart for Chicken Tenders

	Mg needed per gram of Protein	Bread Cr. 9 g	Recipe .3 g	Eggs, whole	Recipe 18.9 g	Flour AP	Recipe 12.9	Parm Regg	Recipe 10 g	Chicken Breast	Recipe	Total Gram	Total Amin o Acid	Total Protein/g
Essential Amino Acids		p/g	Recipe	p/g	Recipe	p/g	Recipe	p/g	Recipe	p/g	Recipe			
Histidine	18	22.22	199.98	24.60	155.00	22.33	288.00	38.70	387.00	31.04	6313.54	232.60	7343.52	31.57
Isoleuciine	25	40.76	366.84	53.33	336.00	34.57	446.00	53.00	530.00	52.74	10727.32		12406.1	53.34
Methionine	25	17.43	156.87	30.16	190.00	17.75	229.00	26.80	268.00	27.68	5630.11		6473.98	27.83
Leucine	55	72.22	649.98	86.35	544.00	68.84	888.00	96.70	967.00	75.00	15255.00		18303.9	78.69
Lysine	51	32.22	289.98	72.54	457.00	22.09	285.00	92.60	926.00	84.94	17276.80		19234.7	82.69
Phenylalanine	47	49.17	442.53	54.13	341.00	50.39	650.00	53.80	538.00	39.63	8060.74		10032.2	43.13
Threonine	27	32.10	288.90	44.13	278.00	27.21	351.00	36.90	369.00	42.20	8583.48		9870.38	42.43
Tryptophan	7	12.15	109.35	13.25	83.50	12.33	159.00	13.50	135.00	11.71	2381.81		2868.66	12.33
Valine	32	45.00	405.00	68.25	430.00	40.23	519.00	68.70	687.00	49.57	10082.54		12123.5	52.12
Amount per recipe		1 1/2 cups		2		1 Cup		1 ounc e		2 lbs.				
Protein per recipe (g)		9		6.30		12.9		10		203.4				

*p/g = protein/gram

Total recipe Protein (g)	# of servings	Protein / serving (g)	Complete Protein
241.60	8.00	30.20	

Barbecue Beef - Crock Pot

This is another very simple recipe, and a way to make use of the BBQ sauce recipe provided earlier. I find the best way to convince people to take the time to make their own base ingredients, such as BBQ sauce, hot sauce, aioli, is to provide several recipes that use them.

By using our low salt and little or no tomato BBQ sauce, we are reducing both the sodium and potassium in this dish. I also like the idea of using a 3 1/2 pound roast and breaking the leftovers into portions for later use. Having items ready to thaw and eat not only saves time for busy people, it also makes it easier for CKD patients to eat healthy when their energy levels and motivations have disappeared.

When making recipes that use beef, using 100% grass fed, pasture raised beef is the best option for your health. I have provided a more detailed explanation in the last section of this book entitled *Food Groups, Ingredients, and Other Important Factors.*

The primary issue for beef is the feed. Food effects our bodies and our health and this is no different for livestock. The food we eat is affected by what is fed to that food. This may not be the most appetizing idea for many to think about, but it applies to all the food we ingest, whether carnivore, herbivore, or omnivore.

If availability or price poses a problem, find the best graded meat available. Since this is a long slow process, the cheaper cuts of meat work best, such as chuck or round,

Food Note

Beef grading is a way to determine the quality of the meat you are buying. The USDA created the system and it is a completely voluntary program. The factors that go into the grading system are aging and marbling of fat. From these the producer can determine the quality of the meat based on tenderness and hence flavor. Prime, Choice, and Select are the first three in that oder. There are other grades below those that are used for store brand selections, ground beef, and processed food.

Ingredients:

- 3 1/2 pound chuck roast
- 2 1/2 cups homemade BBQ sauce
- 1 cup low sodium beef stock (or homemade)
- 2 teaspoons salt
- 1 teaspoon pepper

Place the chuck roast and the beef stock in the slow cooker. Rub 1 cup of the BBQ sauce over the meat and cook for 4 hours on a high setting or until the meat in fork tender and can be easily pulled apart using forks.

When done, remove the beef and place on a cutting board or a wide enough container to easily shred the beef. Add the remaining 1 1/2 cups of BBQ sauce and up to 1 1/2 cups of the cooking liquid. This liquid is made of beef stock, BBQ sauce, and fat that has been cooked off the roast. Use just enough to make the shredded beef moist.

You can serve it as is or as a sandwich, just don't forget to choose bread that is on your list of acceptable foods. This currently means white flour based bread and no whole wheat or other grain flours as they are higher is phosphorus. I also like to add some homemade coleslaw to my meal, either on the side or on the sandwich.

Time: 4 hours Yield: 72 oz. Portions: 24 Serving size: 3 oz.

Food/Health Note

In the cooking process for beef, there is a loss of weight during cooking. The cut of meat, quality, and cooking process all play a roll in the final nutritional outcome. The USDA has provided data using these variables and the charts are available online at https://www.ars.usda.gov/ARSUserFiles/80400525/data/retn/usda_cookingyields_meatpoultry.pdf.
I use these and other charts to verify the information provided.

BBQ Beef

		Fat (g)	Sat. Fat (g)	Phos (mg)	Pot (mg)	Sodium (mg)	Calorie	Carb (g)	Vit A (iu)	Vit C (mg)	Vit K (Mcg)	Vit E (mg) AT	Vit B6 (mg)
		65	**20**	**700**	**3300**	**2300**	**2000**	**275**	**5000**	**60**	**120**	**20**	**2**
Ingredients	**Amt**												
Beef Chuck - Braised	**3.5 lbs**	100.80	38.15	3412.50	4459.00	906.50	3255.00	0.00	0.00	0.00	23.80	7.00	5.25
		155.08%	190.75%	487.50%	135.12%	39.41%	162.75%	0.00%	0.00%	0.00%	19.83%	35.00%	262.50%
Beef Stock - No sodium	**1 cup**	0.00	0.00	223.60	1110.00	0.00	77.50	7.50	0.00	0.00	0.50	0.00	0.25
		0.00%	0.00%	31.94%	33.64%	0.00%	3.88%	2.73%	0.00%	0.00%	0.42%	0.00%	12.50%
BBQ Sauce - Homemade	**2 1/2 cups**	61.80	4.40	289.00	1893.40	173.42	1362.00	135.00	9428.60	161.00	87.00	14.20	1.80
		95.08%	22.00%	41.29%	57.38%	7.54%	68.10%	49.09%	188.57%	268.33	72.50%	71.00%	90.00%
Salt	**2 Tsp**					4600.00							
						200.00%							
Pepper	**1/2 tsp**	0.05	0.00	1.75	12.60	0.45	2.55	0.50	0.03	0.20	1.65	0.00	0.00
		0.08%	0.00%	0.25%	0.38%	0.02%	0.13%	0.18%	0.00%	0.33%	1.38%	0.00%	0.00%
Total for Dish		162.65	42.55	3926.85	7475.00	5680.37	4697.05	143.00	9428.63	161.20	112.95	21.20	7.30
		250.23%	212.75%	560.98%	226.52%	246.97%	234.85%	52.00%	188.57%	268.67	94.13%	106.00%	365.00%
Servings (3 oz)		24	24	24	24	24	24	24	24	24	24	24	24
Totals per serving		6.78	1.77	163.62	311.46	236.68	195.71	5.96	392.86	6.72	4.71	0.88	0.30
		10.43%	8.86%	23.37%	9.44%	10.29%	9.79%	2.17%	7.86%	11.19%	3.92%	4.42%	15.21%

Essential Amino Acid Chart Beef BBQ

	Mg/ gram protein needed	Beef Chuck Braised	Recipe 549.5 g	BBQ Sauce	Recipe 4.73 g	Beef Stock	Recipe 2.25 g	Total Grams	Total Amino Acid	Total Protein/g
Essential Amino Acids		**Protein/g**	**Recipe**	**Protein/g**	**Recipe**	**Protein/g**	**Recipe**			
Histidine	**18**	31.89	17523.56	14.00	66.23		0.00	556.26	17589.78	31.62
Isoleuciine	**25**	45.47	24985.77	18.63	88.12		0.00		25073.88	45.08
Methionine	**25**	26.04	14308.98	5.38	25.46		0.00		14334.44	25.77
Leucine	**55**	79.52	43696.24	30.04	142.07		0.00		43838.31	78.81
Lysine	**51**	84.46	46410.77	32.73	154.84		0.00		46565.61	83.71
Phenylalanine	**47**	39.49	21699.76	24.72	116.94		0.00		21816.69	39.22
Threonine	**27**	39.92	21936.04	22.57	106.74		0.00		22042.78	39.63
Tryptophan	**7**	6.56	3604.72	9.46	44.77		0.00		3649.49	6.56
Valine	**32**	49.60	27255.20	24.29	114.88		0.00		27370.08	49.20
Recipe Amout		3.5 lbs.		2 1/2 cups		2 Tbps				
Protein per recipe (g)		549.5		4.73		2.025				

Total recipe Protein (g)	**# of servings**	**Protein / serving (g)**	Complete Protein
556.26	24.00	23.18	

Buffalo Chicken Wings and Sauce

Chicken wings are not usually a food item you find listed in a nutrition and health guide, at least not in a positive manner. They are usually fried, loaded with a salty, fat laden hot sauce, and served with another fattening dip of blue cheese or ranch dressing.

But they are delicious. So how can I find anything positive for anyone, especially for people with chronic kidney disease? Let's start with who this isn't good for. Most patients in the beginning stages of kidney failure are told to reduce protein intake, and the high protein in meats makes this a less than optimal choice.

The higher fat content is also a factor. Anyone who is at the high end of cholesterol or considered an at risk patient on a reduced fat diet should check with their health care team about this recipe.

O.K., now on to the positives. These wings are baked in the oven, reducing additional fat from frying. Using my homemade hot sauce reduces the sodium levels in the final product. You may recall from another recipe in this book that store bought hot sauces can be loaded with sodium. Hot sauce is a primary ingredient in Buffalo wing sauce, along with a fat, usually butter or a butter substitute.

For those on dialysis that find it difficult to keep their protein levels at suggested levels, this recipe of four wings offers 38.5 grams of protein. In addition the phosphorus to protein level is 5.78 for the finished product. The phosphorus to protein chart in the back of the book shows that this is one of the lowest levels for all the foods listed (lower numbers are better).

The final positive factor addresses the psychological more than the nutritional level. As mentioned earlier, during my time on dialysis, a few things became very obvious. I was going to eat things I should not, and others in that room were doing the same. It was not something that needed any research since patients brought food into the dialysis room. Fast food, bags of salty chips, cookies, crackers, and other items. The staff would remind them of the health implications, and the response, often through a look, said, "Really, you want to take that away from me too?"

My point is that nutrition should not only address the items patients should eat, but also the items we are going to eat. The goal is not to merely tell patients what to eat, but assist in adjusting the foods that are currently in their food rotation

Buffalo Wing Sauce

Chicken wing sauce has two main ingredients, plus additional flavorings for added heat and/or variations in taste. A pepper sauce of your choice and butter are the primary forces, and in this recipe I add a dash of mustard, Worcestershire sauce, and lime juice. That's all you really need. The heat level of the sauce will be similar to the heat level of the pepper sauce, but can be increased by adding crushed red pepper flakes, cayenne pepper, or mixing pepper sauces with greater heat levels.

I make the wing sauce first and cool it down long enough so that it is somewhere between a thin soupy liquid when hot, and the hard firmness of butter when cold. I have found that the sauce sticks to the wings much better when at this mid point of liquid and solid.

Ingredients:

- 1 cup homemade hot pepper sauce (see recipe) or a low salt store bought brand of your choice
- 6 tablespoons unsalted butter
- 1 tablespoon Worcestershire sauce
- 1 teaspoon of mustard or 1/2 teaspoon of mustard powder
- Juice and zest of one lime

Total time: 10 minutes Yield: 1 1/2 cups Portions: 6
Serving size: 1/4 cup

Melt the butter on low heat in a sauce pan and add the remaining ingredients. Mix thoroughly and let cool at room temperature or in the refrigerator.

Food Note

This seems like a good point talk about butter. The first thing to know is never buy salted butter. It is important to restrict sodium in your diet as a CKD patient, and you can control the amount of salt buy starting with ingredients with no or low sodium. The same rule apples as a chef, I want as much control as possible over the final dish.

Butter also has an expiration date. You can still use it past this date, but the flavor starts to deteriorate and becomes rancid after long periods of time. Added salt will slow this process.

Buffalo Wing Sauce

		Fat (g)	Sat Fat (g)	Phos (mg)	Pot (mg)	Sodium (mg)	Calories	Carbs (g)	Vit A (iu)	Vit C (mg)	Vit K (Mcg)	Vit E (mg) AT	Vit B6 (mg)
		65	20	700	3300	2300	2000	275	5000	60	120	20	2
Ingredients	Amt												
Homemade Hot Pepper Sauce	1 cup	0.00	0.00	127.77	999.61	819.20	150.11	23.67	9349.42	341.64	38.39	1.93	1.24
		0.00%	0.00%	18.25%	30.29%	35.62%	7.51%	8.61%	186.99%	569.41%	31.99%	9.67%	62.22%
Butter, unsalted	6 Tbsp	66	42	20.4	20.4	9	600	0	2100	0	6	1.8	0
		101.54'	210.00%	86.00%	0.62%	0.39%	30.00%	0.00%	42.00%	64.80%	5.00%	9.00%	0.00%
Lime Juice/ Zest	1 oz	0.00	0.00	8.00	32.80	1.20	14.00	4.00	14.00	8.40	0.20	0.10	0.00
(2 Tbsp)		0.00%	0.00%	1.14%	0.99%	0.05%	0.70%	1.45%	0.28%	14.00%	0.17%	0.50%	0.00%
Worcestershire	1 Tbsp	0.00	0.00	10.20	136.00	167.00	13.30	3.30	13.40	2.20	0.20	0.00	0.00
		0.00%	0.00%	1.46%	4.12%	7.26%	0.67%	1.20%	0.27%	3.67%	0.17%	0.00%	0.00%
Mustard	1 Tbsp	0.90	0.60	15.90	20.70	170.50	52.00	9.90	10.80	0.30	0.30	0.00	0.00
		1.38%	3.00%	2.27%	0.63%	7.41%	2.60%	3.60%	0.22%	0.50%	0.25%	0.00%	0.00%
Total for Dish		66.90	42.60	182.27	1209.51	1166.90	829.41	40.87	11487.62	352.54	45.09	3.83	1.24
		102.92'	213.00%	26.04%	36.65%	50.73%	41.47%	14.86%	229.75%	587.57%	37.57%	19.17%	62.22%
# of Servings		6	6	6	6	6	6	6	6	6	6	6	6
Totals per serving		11.15	7.10	30.38	201.59	194.48	138.24	6.81	1914.60	58.76	7.51	0.64	0.21
		17.15%	35.50%	4.34%	6.11%	8.46%	6.91%	2.48%	38.29%	97.93%	6.26%	3.19%	10.37%

Total recipe Protein (g)	# of servings	Protein / serving (g)	Not a Complete Protein
3.23	6.00	0.54	

Essential Amino Acid Chart for recipe Buffalo Wing Sauce

	Mg/ gram protein needed	Hot pepper sauce	Recipe	Butter	Recipe .6 g	Lime - Juice & Zest	Recipe 10 g	Total Grams	Total Amino Acid	Total Protein/ g
Essential Amino Acids		Protein/ g	Recipe	Protein/g	Recipe	Protein/g	Recipe			
Histidine	18	21.58	54.59	27.47	16.48	38.70	3.87	3.23	74.95	23.20
Isoleuciine	25	34.21	86.55	61.05	36.63	53.00	5.30		128.48	39.78
Methionine	25	12.63	31.96	25.11	15.06	26.80	2.68		49.70	15.39
Leucine	55	55.26	139.82	98.95	59.37	96.70	9.67		208.85	64.66
Lysine	51	46.84	118.51	80.00	48.00	92.58	9.26		175.77	54.42
Phenylalanine	47	32.63	82.56	49.00	29.40	53.80	5.38		117.34	36.33
Threonine	27	38.95	98.54	45.42	27.25	36.90	3.69		129.48	40.09
Tryptophan	7	13.68	34.62	14.32	8.59	13.50	1.35		44.56	13.80
Valine	32	44.21	111.85	67.89	40.74	68.70	6.87		159.46	49.37
Recipe Amout		1 cup		6 Tbsp		1 each				
Protein per recipe (g)		2.53		0.6		0.1				

Baked Chicken Wings

Ingredients:

- 2 dozen chicken wings, thawed and patted dry
- 2 cups flour
- 2 tablespoons paprika
- 1 teaspoon oregano
- 1 teaspoon cayenne pepper
- 1 teaspoon crushed red pepper
- 1 1/2 cups homemade wing sauce

Time: 60 minutes Yield: 2 dozen wings Portions: 6
Serving size: 4 wings

Baking chicken wings can often lead to soggy and light colored wings. Frying helps remove water in the skin faster than baking, which makes the wings crisper. There are two methods that work well to get baked wings to look and taste like they are fried. The first is high heat, 450° F or higher. This will crisp the outside of the wing by drying out the skin. The problem here is that the meat of the wings also dries out and the meat is tougher.

The other method is set the wings on a cooling rack placed in a rimmed baking dish, and set uncovered in the refrigerator overnight. This gets the drying out process started and when baked at 400° F, makes the skin crisp and the meat still moist.

You can still bake the wings without the drying process, but there will be more variability in the crisp skin to moist meat ratio.

Heat the oven to 400° F. For the pan I suggest using a wire rack on rimmed sheet pan for cooking. This will increase the surface area exposed to the heat, and allow the fat to drip away from the food. It is also helpful to spray the wire rack with non stick spray so the wings will not stick to the wires.

Combine the flour and all the spices in a bowl. Toss the wings into the flour mixture and place on the wire rack, leaving a little space between each wing. You will have flour left over, and that is why their is difference between the recipe listing and the charts, only half of the flour will be consumed

Food Note

When baking wings, there are two schools of thought. One is to toss the wings in flour before baking, the other uses no flour at all. The flour is thought to add to the browning of the wings, making them crisp. This is called the Maillard reaction. In short, this is the dark crisp exterior on food when cooked at high temperatures. The reaction happens between amino acids and a reducing sugar. By adding salt to meat and letting is sit for 20 minutes to an hour, amino acids get pulled from inside the meat to the surface, helping to increase the reaction. But this is a problem for CKD patients. The drying method in the recipe, removes moisture, and will assist the crisping without adding salt.

In addition to the crispness, using additional spices mixed in the flour will add flavor as the salt levels are lower than standard chicken wings.

Cook the wings for 25 minutes, rotate or roll the wings over, and cook for another 20 minutes, or until they look crisp on the outside.

At this point you can either toss them in the sauce, or refrigerate them for up to 5 days. If you choose to do this, rewarm the wings in the oven for 15 minutes at 400° F and then toss in the wing sauce. Making several batches ahead of time will allow you to serve several dozen wings at once, and it will only take about 20 minutes.

I have found that tossing the wings in a heated sauce reduces it's adherence. A room temperature sauce works best to help it stick to the wings.

Buffalo Chicken Wings

		Fat (g)	Sat. Fat (g)	Phos (mg)	Pot (mg)	Sodium (mg)	Calories	Carb (g)	Vit A (iu)	Vit C (mg)	Vit K (Mcg)	Vit E (mg) AT	Vit B6 (mg)
	RDA	65	20	700	3300	2300	2000	275	5000	60	120	20	2
Ingredients	Amt												
Fresh/Frozen chicken wings	2 dozen	156.0	45.60	615.60	1509.60	669.60	2366.40	0.00	1288.80	0.00	2.40	2.40	2.40
		240.0	228.00%	87.94%	45.75%	29.11%	118.32%	0.00%	25.78%	0.00%	2.00%	12.00%	120.00%
Homemade Wing sauce	1 1/2 cups	66.00	42.00	166.37	1188.81	996.40	777.41	30.97	11476.82	352.24	44.79	3.83	1.24
		101.5	210.00%	23.77%	36.02%	43.32%	38.87%	11.26%	229.54%	587.07%	37.32%	19.17%	62.22%
Flour	1 cup	1.20	0.20	134.00	135.00	2.50	455.00	95.00	2.50	0.00	0.40	0.30	0.10
		1.85%	1.00%	19.14%	4.09%	0.11%	22.75%	34.55%	0.05%	0.00%	0.33%	1.50%	5.00%
Cayenne pepper	1 tsp	0.30	0.10	5.10	35.20	0.50	5.60	1.00	728.00	1.30	1.40	0.50	0.00
		0.46%	0.50%	0.73%	1.07%	0.02%	0.28%	0.36%	14.56%	2.17%	1.17%	2.50%	0.00%
Oregano	1 tsp	0.00	0.00	2.00	16.70	0.10	3.10	1.00	69.00	0.50	6.20	0.20	0.00
		0.00%	0.00%	0.29%	0.51%	0.00%	0.16%	0.36%	1.38%	0.83%	5.17%	1.00%	0.00%
Crushed Red peeper flakes	1 tsp	0.30	0.10	5.10	35.20	0.50	5.60	1.00	728.00	1.30	1.40	0.50	0.00
		0.46%	0.50%	0.73%	1.07%	0.02%	0.28%	0.36%	14.56%	2.17%	1.17%	2.50%	0.00%
Paprika	1 Tbsp	0.90	0.10	23.30	158.00	2.30	19.50	3.80	3560.00	4.80	5.40	2.00	0.30
		1.38%	0.50%	3.33%	4.79%	0.10%	0.98%	1.38%	71.20%	8.00%	4.50%	10.00%	15.00%
Salt	1/4 tsp					575.00							
						25.00%							
Pepper	1/2 tsp	0.05	0.00	1.75	12.60	0.45	2.55	0.50	0.03	0.20	1.65	0.00	0.00
		0.08%	0.00%	0.25%	0.38%	0.02%	0.13%	0.18%	0.00%	0.33%	1.38%	0.00%	0.00%
Total for Dish		224.7	88.10	953.22	3091.11	2247.35	3635.16	133.27	17853.15	360.34	63.64	9.73	4.04
		345.7	440.50%	136.17%	93.67%	97.71%	181.76%	48.46%	357.06%	600.57%	53.03%	48.67%	202.22%
Servings 3 (4 wings)		6	6	6	6	6	6	6	6	6	6	6	6
Totals per serving		37.46	14.68	158.87	515.19	374.56	605.86	22.21	2975.53	60.06	10.61	1.62	0.67
		57.63	73.42%	22.70%	15.61%	16.29%	30.29%	8.08%	59.51%	100.10%	8.84%	8.11%	33.70%

Essential Amino Acid Chart Buffalo Chicken Wings

	Mg/ gram protein needed	Chicken Wings	Recipe 109.2 g	Buffalo wing sauce	Recipe 2.53 g	Flour - All purpose	Recipe 3.23 g	Total Grams	Total Amino Acid	Total Protein/g
Essential Amino Acids		Protein/g	Recipe	Protein/g	Recipe	Protein/g	Recipe			
Histidine	**18**	28.02	3060.00	21.95	55.53	22.33	72.00	114.96	3187.53	27.73
Isoleuciine	**25**	47.47	5184.00	43.41	109.83	34.57	111.50		5405.33	47.02
Methionine	**25**	25.71	2808.00	14.90	37.70	17.75	57.25		2902.95	25.25
Leucine	**55**	70.88	7740.00	63.57	160.83	68.84	222.00		8122.83	70.66
Lysine	**51**	78.46	8568.00	53.05	134.22	22.09	71.25		8773.47	76.32
Phenylalanine	**47**	38.24	4176.00	35.98	91.03	50.39	162.50		4429.53	38.53
Threonine	**27**	40.66	4440.00	40.23	101.77	27.21	87.75		4629.52	40.27
Tryptophan	**7**	10.73	1171.20	13.51	34.19	12.33	39.75		1245.14	10.83
Valine	**32**	47.69	5208.00	50.85	128.64	40.23	129.75		5466.39	47.55
Recipe Amout		24 wings		1 1/2 cup		1 cup				
Protein per recipe (g)		109.2		2.53		3.225				

Total recipe Protein (g)	**# of servings**	**Protein / serving (g)**	Complete Protein
114.96	4.00	28.74	

Baked Zucchini Sticks

This recipe makes for a very good substitute side dish for potatoes. The added flavor in the Parmigiano Reggiano will help reduce the use of sodium, and the panko breadcrumbs adds a nice crunchy texture. Baking rather than frying can reduce the fat content keeping them in a healthier zone. I make a large batch and freeze servings in smaller bags. Warm them in the oven for 15 minutes at 350° F on the days you do not feel like cooking.

Baking is a safer and healthier way to cook most foods, as it is often difficult to limit the additional oils from getting into the food. You can still create a nice crust with a little oven management for many baked items.

Ingredients:

- 2 medium size zucchini
- 2 large eggs
- 1 1/2 cups panko breadcrumbs
- 1/4 cup Parmigiano Reggiano cheese
- 2 tablespoons olive oil
- 1 cup all purpose flour
- 1/2 teaspoon black pepper

Total Time: 45 minutes Yield: 2 cups Portions : 4

Serving size: 1/2 cup

This recipe uses the traditional 3 station breading system. There will be some left over in each station and the data below takes that into account.

Start by turning your oven on to 425° F, and set up your 3 breading stations.

In bowl 1, mix the flour and black pepper together. In bowl 2, whisk the eggs until whites and yolks are combined. In bowl 3, mix the panko breadcrumbs and the Parmigiano Reggiano together.

My preference is large metal bowls with high sides so I can toss the ingredients, and make less of a mess.

Food Note

Baking and frying are considered dry heat forms of cooking. Moist heat cooking methods use a liquid to transfer the heat into food. They include braising, steaming, boiling, and poaching. Dry heat uses air or fat to transfer heat.
That sizzling noise you hear when frying is the water in the food boiling, turning to steam and escaping. It is pushing it's way out of the food, through the oil and into the air. While it is pushing out, the steam creates a barrier around the food item so oil does not penetrate. Once the outside of the food item is cooked and creates a seal, the leftover water in the food boils, and cooks the inside since it cannot escape. As the bubbling slows down and the water is not creating that barrier, oil is getting in. The key to frying is using a proper oil temperature for the particular food. It needs to be hot enough to create a crust on the outside while fully cooking the inside. The goal is to have this happen at the same time to avoid oil seeping into the food, or burning the outside crust.

Now prep the zucchini. Cut each zucchini in half through the round part. Next cut each of the halves lengthwise into the size of large steak fries. You should get about 3 to 4 per section.

If you have a cooling rack that will fit into a sheet pan, that is the best option. If not, a sheet pan alone will work fine, but be sure to spray the pan with a non-stick product of your choice.

Drop the zucchini in the flour and coat. Be sure to let all the excess flour fall off. Next place zucchini in the egg mixture, and let excess drip off. Finally toss into the panko mixture and toss to coat well. Place on cooling rack or pan and repeat.

Be sure to leave a little room between each piece and never overlap. One layer only. Bake for 25 minutes or until the crust starts to brown.

Breaded Zucchini Sticks (Baked)

		Fat (g)	Sat. Fat (g)	Phos (mg)	Pot (mg)	Sodium (mg)	Calories	Carbs (g)	Vit A (iu)	Vit C (mg)	Vit K (Mcg)	Vit E (mg) AT	Vit B6 (mg)
	RDA	65	20	700	3300	2300	2000	275	5000	60	120	20	60
Ingredients	Amt												
Eggs	2 Large	10.00	4.00	63.67	44.67	143.33	71.00	0.00	162.33	0.00	0.10	0.33	0.03
		15.38%	20.00%	9.10%	1.35%	6.23%	3.55%	0.00%	3.25%	0.00%	0.08%	1.65%	0.05%
Panko Bread Crumbs	1 1/2 cups	0	0	0	0	141	156	33	18.612	131.01	7.1	0.1	0.1
		0.00%	0.00%	0.00%	0.00%	6.13%	7.80%	12.00%	0.37%	64.80%	5.92%	0.50%	0.17%
Zucchini	2 med	2	0	94.4	650	24.8	40	8	496	42.2	10.6	0.2	0.6
(2 cups)		3.08%	0.00%	13.49%	19.70%	1.08%	2.00%	2.91%	9.92%	70.33%	8.83%	1.00%	1.00%
Parmigiano Reggiano Fresh	1/4 cup (1 oz)	3	5	194	25.8	449	40	2	112	0	0.5	0.1	0
(2 g)		4.62%	25.00%	27.71%	0.78%	19.52%	2.00%	0.73%	2.24%	0.00%	0.42%	0.50%	0.00%
Olive Oil	1 Tbsp	13.50	1.88	0.20	0.40	0.00	179.09	0.00	0.00	0.00	12.19	2.91	0.00
(2 g)		20.77%	9.38%	0.03%	0.01%	0.00%	8.95%	0.00%	0.00%	0.00%	10.16%	14.53%	0.00%
Flour, White AP Enriched.	1 cup Use 1/2	0.5	0.1	67.5	67	1	227.5	23.75	0	0	0.2	0.15	0.05
(6 oz. or 168 g)		0.77%	0.50%	9.64%	2.03%	0.04%	11.38%	8.64%	0.00%	0.00%	0.17%	0.75%	2.00%
Total for Dish		29.00	10.98	419.77	787.87	759.13	713.59	66.75	788.95	173.21	30.69	3.79	0.78
		44.62%	54.88%	59.97%	23.87%	33.01%	35.68%	24.27%	15.78%	288.68%	25.58%	18.93%	1.30%
# of Servings	4 oz. ser.	4	4	4	4	4	4	4	4	4	4	4	4
Totals per serving	4 Oz	7.25	2.74	104.94	196.97	189.78	178.40	16.69	197.24	43.30	7.67	0.95	0.20
		11.15%	13.72%	14.99%	5.97%	8.25%	8.92%	6.07%	3.94%	72.17%	6.39%	4.73%	0.33%

Essential Amino Acid Chart for recipe Baked Zucchini Sticks

	Mg needed per gram of Protein	Bread Cr. panko	Recipe 12 g	Eggs, whole	Recipe 18.9 g	Flour Ap	Recipe 12.9	Parmigiano Reggiano,	Recipe 10 g	Zucchini	Recipe 3 g	Total Gram	Total Amino Acid	Total P/g
Essential Amino Acids		p/g	Recipe	p/g	Recipe	p/g	Recipe	p/g	Recipe	p/g	Recipe			
Histidine	18	22.22	266.64	24.60	155.00	22.33	145.12	38.70	387.00	21.67	65.01	25.80	1018.77	39.49
Isoleuciine	25	40.76	489.12	53.33	336.00	34.57	224.73	53.00	530.00	36.40	109.20		1689.05	65.47
Methionine	25	17.43	209.16	30.16	190.00	17.75	115.39	26.80	268.00	14.87	44.62		827.17	32.06
Leucine	55	72.22	866.64	86.35	544.00	68.84	447.44	96.70	967.00	58.76	176.29		3001.37	116.33
Lysine	51	32.22	386.64	72.54	457.00	22.09	143.60	92.60	926.00	55.43	166.28		2079.53	80.60
Phenylalanine	47	49.17	590.04	54.13	341.00	50.39	327.52	53.80	538.00	35.55	106.65		1903.21	73.77
Threonine	27	32.10	385.20	44.13	278.00	27.21	176.86	36.90	369.00	24.01	72.04		1281.10	49.65
Tryptophan	7	12.15	145.80	13.25	83.50	12.33	80.12	13.50	135.00	8.27	24.81		469.23	18.19
Valine	32	45.00	540.00	68.25	430.00	40.23	261.51	68.70	687.00	44.69	134.07		2052.58	79.56
Amount per recipe		1 cup		2		1 Cup		1 ounce		2 med= 2 cup				
Protein per recipe (g)		12		6.30		6.5		10		3				

Total recipe Protein (g)	# of servings	Protein / serving (g)	Not a Complete Protein
37.80	4.00	9.45	

1/2 CUP

Lemon-Lime Simple Syrup

One of the first items CKD patients (and frankly everyone) should avoid are most varieties of pre-made soft drinks. Whether you call them soda, pop, soda pop, or "Coke", these canned and bottled items are generally just a bad idea. Between the corn syrup, sodium, and cola beans, nothing here is going to be helpful for people with kidney issues.

Given that information, it certainly does not stop people from consuming them. Sometimes it is the caffeine, sometimes the refreshing feel of the cold bubbles and flavor on a hot day, and often is just a behavior we have developed and are unwilling to omit from our diets.

This recipe is a way to lessen some of the issues from those products and a great way to avoid some of the sugar and most of the sodium in those pre-made products.

The flavor combinations are endless, and you can almost always find a way to satisfy your taste. The exception might be for cola products as it is the cola bean that contains the high levels of phosphorus.

Simple syrup is a sweetener, and can also add flavor to soft drinks. It is used in many cocktails, if that is something acceptable to you and your care team.

It is very easy to make, and will last in the fridge for several weeks. I have suggested a portion size of 2 ounces, but you may not need that much for your taste.

Ingredients:

- 2 cups water
- 1 1/2 cups sugar
- Juice and zest from 1 lemon
- Juice and zest from 1 lime

Total time: 15 minutes Yield: 2 cups Portions: 8
Serving size: 4 oz.

Place all the ingredients into a small pot on medium heat. Stir occasionally until all the sugar is dissolved. Take off the heat and cool or place directly in a container and refrigerate.

This is where that microplane comes in handy, if you have one. I prefer to grate the peel of the lemon and lime and leave the little pieces in the simple syrup. If you choose this option, remember not to grate into the pith - the part between the rind and the edible part of the citrus. It is white in color, very bitter, and unpleasant.

You can also use a peeler to cut off several pieces of the rind rather than using the microplane. Add these strips while heating and let sit in the simple syrup to increase the flavor. It will also give you a clear end product, with just a little tint of the fruit color.

Health Note

Fluid intake for CKD patients in stages 1 though 4 depends on a several factors, all of which should be discussed with your nephrologist. An older suggestion from primary care physicians (PCP's) was to increase fluid during these stages for a variety of different reasons, most of which have not been proven, with the exception of a few particular types of kidney disease (Wenzel, Herbert, Stahl, & Krenz, 2006). In addition to the cause of your kidney disease, other medical issues and your personal health will play a role.

If you reach stage 5 or ESRD, you are placed on dialysis. This means your body is not removing enough toxins and fluid for you to survive. Dialysis removes both, but between treatments your body will carry the extra fluid and that can be very dangerous. The article "Fluid overload in dialysis patients" (2018) on the NKF website tells us the extra fluid causes swelling, discomfort, high blood pressure, shortness of breath, and heart problems.

Lemon/Lime Simple Syrup

		Fat (g)	Sat. Fat (g)	Phos (mg)	Pot (mg)	Sodium (mg)	Calories	Carb (g)	Vit A (iu)	Vit C (mg)	Vit K (Mcg)	Vit E (mg) AT	Vit B6 (mg)
	RDI	**65**	**20**	**700**	**3300**	**2300**	**2000**	**275**	**5000**	**60**	**120**	**20**	**2**
Ingredients	**Amt**												
Water	**1 cup**	0.00	0.00	0.00	0.00	0.00	0.00	0.00	0.00	0.00	0.00	0.00	0.00
		0.00%	0.00%	0.00%	0.00%	0.00%	0.00%	0.00%	0.00%	0.00%	0.00%	0.00%	0.00%
Sugar	**3/4 cup**	0	0	0	0	0	580.5	150	0	0	0	0	0
		0.00%	0.00%	0.00%	0.00%	0.00%	29.03%	54.55%	0.00%	64.80%	0.00%	0.00%	0.00%
Lemon Juice	**1 oz.**	0.00	0.00	13.80	36.10	1.50	13.00	3.00	0.80	2.80	0.20	0.00	0.10
		0.00%	0.00%	3.99%	1.09%	0.60%	0.65%	1.09%	0.02%	4.67%	0.17%	0.00%	5.00%
Lime Juice	**1 oz.**	0.00	0.00	3.99	32.80	0.60	7.00	2.00	14.00	8.40	0.20	0.10	0.00
		0.00%	0.00%	0.57%	0.99%	0.03%	0.35%	0.73%	0.28%	14.00%	0.17%	0.50%	0.00%
Total for Dish		0.00	0.00	17.79	68.90	2.10	600.50	155.00	14.80	11.20	0.40	0.10	0.10
		0.00%	0.00%	2.54%	2.09%	0.09%	30.03%	56.36%	0.30%	18.67%	0.33%	0.50%	5.00%
# of Servings		8	8	8	8	8	8	8	8	8	8	8	8
(2 Tbsp)													
Totals per serving		0.00	0.00	2.22	8.61	0.26	75.06	19.38	1.85	1.40	0.05	0.01	0.01
		0.00%	0.00%	0.32%	0.26%	0.01%	3.75%	7.05%	0.04%	2.33%	0.04%	0.06%	0.63%

Essential Amino Acid Chart for recipe Lemon/Lime Simple Syrup

	Mg/ gram protein needed	Sugar	Recipe 0 g	Lemon Juice/ Zest	Recipe .2 g	Lime Juice/zest	Recipe .2 g	Total Grams	Total Amino Acid	Total Protein/g
Essential Amino Acids		**Protein/g**	**Recipe**	**Protein/g**	**Recipe**	**Protein/g**	**Recipe**			
Histidine	**18**	0.00	0.00	6.00	1.20	6.00	1.20	0.40	2.40	6.00
Isoleuciine	**25**	0.00	0.00	6.00	1.20	6.00	1.20		2.40	6.00
Methionine	**25**	0.00	0.00	6.00	1.20	6.00	1.20		2.40	6.00
Leucine	**55**	0.00	0.00	45.00	9.00	45.00	9.00		18.00	45.00
Lysine	**51**	0.00	0.00	45.00	9.00	45.00	9.00		18.00	45.00
Phenylalanine	**47**	0.00	0.00	31.00	6.20	31.00	6.20		12.40	31.00
Threonine	**27**	0.00	0.00	6.00	1.20	6.00	1.20		2.40	6.00
Tryptophan	**7**	0.00	0.00	6.00	1.20	6.00	1.20		2.40	6.00
Valine	**32**	0.00	0.00	31.00	6.20	31.00	6.20		12.40	31.00
Recipe Amount		1 Cup		2 oz.		2 oz.				
Protein per recipe (g)		0		0.2		0.2				

Total recipe Protein (g)	# of servings	Protein / serving (g)	Not a Complete Protein
0.40	6.00	0.07	

Lemonade - Limeade

Now that you have a simple syrup, or a few of them in the fridge, it's time to make some refreshing drinks.

These are all pretty easy, however, you should always check with your health care team about acceptable amounts and types of fluid intake. Not all of these will be acceptable for everyone with CKD.

First let's make a simple and easy lemonade. Once you have the simple syrup, all you need is some fresh lemons.

Ingredients:

- 2 ounces lemon/lime simple syrup (recipe above)
- 3 ounces fresh lemon juice (1 1/2 lemons juiced)
- 8 ounces water or soda water

You can adjust the recipe to meet your tastes. Add more lemon juice for tartness; more simple syrup for sweetness; and more water to lighten up the flavors.

Replace the lemon juice above with lime juice, or a combination of the two, for a drink similar to the lemon/lime brand drinks. All are very refreshing and will keep you away from the bottled sodas which tend to be high in calories and sodium.

The nutrition data is the same when using soda water for a bubbly lemonade. However, if you use tonic water you will be adding about 30 grams of sugar and an 80 calories per 12 oz. I would suggest staying away from tonic as there is plenty of sugar in the simple syrup.

Food Science Note

Although waters may appear to be very similar, there are some significant differences. Seltzer, or soda water, is made by adding carbonation to plain water. Tonic water is made by adding sugar at high levels and quinine at low levels, along with carbonation. Originally used as a cure for malaria, quinine in tonic water is about 10% of the dose for malaria cures. It does have quite a lot of sugar added, and therefore is loaded with carbohydrates.

Even tap water has many different characteristics that should be monitored. You can do this by contacting your local water authority. They have a list of the different minerals, and the amounts, in the water source to your house.

Fluoride is a common mineral found in tap water. If you have kidney issues, high levels of fluoride can reduce the absorption of magnesium in the body (Machoy-Mokrzynska, n.d.). Research by Eby and Eby (2006) shows that low magnesium levels are associated with major depression and supplements are being offered as treatment. As always, check with your care professionals before making any adjustments to your diet.

Lemon/Lime -ade

		Fat (g)	Sat. Fat (g)	Phos (mg)	Pot (mg)	Sodium (mg)	Calories	Carb (g)	Vit A (iu)	Vit C (mg)	Vit K (Mcg)	Vit E (mg) AT	Vit B6 (mg)
	RDA	**65**	**20**	**700**	**3300**	**2300**	**2000**	**275**	**5000**	**60**	**120**	**20**	**2**
Ingredients	**Amt**												
Water	**8 oz**	0.00	0.00	0.00	0.00	0.00	0.00	0.00	0.00	0.00	0.00	0.00	0.00
		0.00%	0.00%	0.00%	0.00%	0.00%	0.00%	0.00%	0.00%	0.00%	0.00%	0.00%	0.00%
Lemon/Lime Simple Syrup	**4 oz**	0.00	0.00	5.94	22.96	0.70	200.16	51.66	4.94	3.74	0.14	0.04	0.06
		0.00%	0.00%	0.85%	0.70%	0.03%	10.01%	18.79%	0.10%	64.80%	0.12%	0.20%	3.00%
Fresh Lemon Juice and zest	**4 oz**	0.00	0.00	27.60	72.20	3.00	26.00	6.00	1.60	5.60	0.40	0.00	0.20
		0.00%	0.00%	3.99%	2.19%	0.60%	1.30%	2.18%	0.03%	9.33%	0.33%	0.00%	10.00%
Fresh Lime Juice and zest	**4 oz**	0.00	0.00	7.98	64.60	1.20	14.00	4.00	28.00	16.80	0.40	0.20	0.00
		0.00%	0.00%	1.14%	1.96%	0.05%	0.70%	1.45%	0.56%	28.00%	0.33%	1.00%	0.00%
Total for Dish		0.00	0.00	41.52	159.76	4.90	240.16	61.66	34.54	26.14	0.94	0.24	0.26
		0.00%	0.00%	5.93%	4.84%	0.21%	12.01%	22.42%	0.69%	43.57%	0.78%	1.20%	13.00%
# of Servings		1	1	1	1	1	1	1	1	1	1	1	1
Totals per serving		0.00	0.00	41.52	159.76	4.90	240.16	61.66	34.54	26.14	0.94	0.24	0.26
		0.00%	0.00%	5.93%	4.84%	0.21%	12.01%	22.42%	0.69%	43.57%	0.78%	1.20%	13.00%

Essential Amino Acid Chart for Lemon/Limeade

Essential Amino Acids	Mg/ gram protein needed	Lemon/ Lime Simple Syup	Recipe . 2 g	Lemon Juice/Zest	Recipe .1 g	Lime Juice/ Zest	Recipe . 1 g	Total Grams	Total Amino Acid	Total Protein/g
		Protein/g	Recipe	Protein/g	Recipe	Protein/g	Recipe			
Histidine	18	6.00	1.20	6.00	0.60	6.00	0.60	0.40	1.20	3.00
Isoleuciine	25	6.00	1.20	6.00	0.60	6.00	0.60		1.20	3.00
Methionine	25	6.00	1.20	6.00	0.60	6.00	0.60		1.20	3.00
Leucine	55	45.00	9.00	45.00	4.50	45.00	4.50		9.00	22.50
Lysine	51	45.00	9.00	45.00	4.50	45.00	4.50		9.00	22.50
Phenylalanine	47	31.00	6.20	31.00	3.10	31.00	3.10		6.20	15.50
Threonine	27	6.00	1.20	6.00	0.60	6.00	0.60		1.20	3.00
Tryptophan	7	6.00	1.20	6.00	0.60	6.00	0.60		1.20	3.00
Valine	32	31.00	6.20	31.00	3.10	31.00	3.10		6.20	15.50
Recipe Amout	1 drink									
Protein per recipe (g)		0.2		0.1		0.1				

Total recipe Protein (g)	# of servings	Protein / serving (g)	Not a Complete Protein
0.40	1.00	0.40	

Berry Shrubs - The Drinks

Shrubs are a very old, and now new again, mixture added to other liquids for flavor. They are a system of preserving fruit (or vegetables) in a vinegar and sugar mixture. Spices are often added to make flavors more complex and interesting. Historically, shrubs can be traced back to at least the days of the Roman Empire. They were used by the military to help keep them hydrated and to quench thirst. This is not something you drink on it's own, but a flavor additive to other drinks. It creates a sweet and sour delight for alcoholic and non-alcoholic drinks.

> Health Note
>
> Vinegar is a natural stimulant to the salivary glands and makes your mouth water. When this happens you tend to drink less and not feel as thirsty. The stated benefits of apple cider vinegar are too numerous to even try to list here. Johnston and Gaas (2006) found some interesting results including indications that it can aid in blood glucose levels and systolic blood pressure. You can read the studies yourself, but few suggest any harm when ingested at low levels.

This shrub is made with berries and is great for quenching thirst with a sweet tangy flavor. It takes a few days for the vinegar flavor to mellow and the fruit and sugar to be at the forefront of the flavor profile, so be patient.

Ingredients:

- 1 cup fresh blackberries
- 1 cup fresh raspberries
- 3/4 cup sugar
- 1 1/4 cup apple cider vinegar

Total prep time: 20 minuters

Total time: 4 days Yield: 1 1/2 cups Portions: 24
Serving size: 1 Tbsp

Combine all the berries and the sugar in a bowl and macerate the mixture to a pulp-like consistency. Cover and place in the fridge for one day.

The next day, mash the mixture through a sieve or fine mesh strainer. Add the vinegar to the syrup mixture, cover, and place back in fridge for three more days. Your Shrub is now ready to serve. This is where the fun starts. The combinations are endless.

Find the balance of fruit and vegetable flavors you enjoy together and create many wonderful cocktails. Your shrub will last at least 3 weeks in the fridge, as these were originally meant to last longer than that. Be smart, if it starts to develop a foul odor, do not use and discard.

This is the cold version for making shrubs and there is also a quicker hot version. Place your fruit or vegetables in a small pot with sugar and a little water. Simmer until the mixture creates a nice syrup. Drain the mixture through a sieve, and then add your vinegar to the syrup. Place in fridge for a day or two before you start using it as it will mellow with time.

DISCLAIMER: I have no idea if you should be drinking alcohol. Talk to your doctor. What I do know is that many patients do drink, with the exception of the mormon woman I sat next to during dialysis for years. So if you choose to, and I'm not saying it's OK to do so, these options can reduce overall fluid intake.

In the 19th century, shrubs found their way into cognacs and bourbons. You can use any liquor to spice up your shrub. I have seen many combinations at new gastro-pubs or cocktail bars. The Berry Shrub goes particularly well with gin or vodka. But please, as I am a big advocate of eating well, I also urge you to drink well. Choose a quality brand for your cocktails. Your body will thank you.

2 Berry Shrub

		Fat (g)	Sat. Fat (g)	Phos (mg)	Pot (mg)	Sodium (mg)	Calories	Carb (g)	Vit A (iu)	Vit C (mg)	Vit K (Mcg)	Vit E (mg) AT	Vit B6 (mg)
	RDA	**65**	**20**	**700**	**3300**	**2300**	**2000**	**250**	**5000**	**60**	**120**	**20**	**2**
Ingredients	**Amt**												
Apple Cider Vinegar	**1 1/4 cup**	0.00	0.00	23.88	342.50	15.00	62.50	2.50	0.00	0.00	0.00	0.00	0.00
		0.00%	0.00%	3.41%	10.38%	0.65%	3.13%	1.00%	0.00%	0.00%	0.00%	0.00%	0.00%
Sugar	**3/4 cup**	0	0	0	0	0	580.5	500	4200	0	0.1	0.4	0
		0.00%	0.00%	0.00%	0.00%	0.00%	54.00%		84.00%	64.80%	0.08%	2.00%	0.00%
Fresh Rasberries	**1 cup**	1.70	0.00	27.10	244.80	1.20	89.80	15.00	117.00	32.20	9.60	0.00	0.10
		2.62%	0.00%	3.99%	7.42%	0.60%	4.49%	6.00%	2.34%	53.67%	8.00%	0.00%	5.00%
Fresh Blackberries	**1 cup**	1.30	0.00	27.40	305.50	1.50	83.50	18.60	288.00	30.24	28.50	5.00	0.07
		2.00%	0.00%	3.91%	9.26%	0.07%	4.18%	7.44%	5.76%	50.40%	23.75%	25.00%	3.50%
Total for Dish		3.00	0.00	78.38	892.80	17.70	816.30	536.10	4605.00	62.44	38.20	5.40	0.17
		4.62%	0.00%	11.20%	27.05%	0.77%	40.82%	214.44	92.10%	104.07%	31.83%	27.00%	8.50%
# of Servings		24	24	24	24	24	24	24	24	24	24	24	15
Totals per serving		0.13	0.00	3.27	37.20	0.74	34.01	22.34	191.88	2.60	1.59	0.23	0.01
		0.19%	0.00%	0.47%	1.13%	0.03%	1.70%	8.94%	3.84%	4.34%	1.33%	1.13%	0.57%

Essential Amino Acid Chart for Berry Shrub

	Mg/ gram protein needed	Apple Cider Vinegar	Recipe 0 g	Black-berries	Recipe 1.7 g	Rasp-berries	Recipe 1.8 g	Total Grams	Total Amino Acid	Total Protein/g
Essential Amino Acids		Protein/g	Recipe	Protein/g	Recipe	Protein/g	Recipe			
Histidine	18	0.00	0.00		0.00	0.00	0.00	3.5	0.00	0.00
Isoleuciine	25	0.00	0.00		0.00	0.00	0.00		0.00	0.00
Methionine	25	0.00	0.00		0.00	0.00	0.00		0.00	0.00
Leucine	55	0.00	0.00		0.00	0.00	0.00		0.00	0.00
Lysine	51	0.00	0.00		0.00	0.00	0.00		0.00	0.00
Phenylalanine	47	0.00	0.00		0.00	0.00	0.00		0.00	0.00
Threonine	27	0.00	0.00		0.00	0.00	0.00		0.00	0.00
Tryptophan	7	0.00	0.00		0.00	0.00	0.00		0.00	0.00
Valine	32	0.00	0.00		0.00	0.00	0.00		0.00	0.00
Recipe Amount		1 Cup		1 cup		1 cup				
Protein per recipe (g)		0		1.7		1.8				

Total recipe Protein (g)	# of servings	Protein / serving (g)	Not a Complete Protein
3.50	24.00	0.15	

Ginger Simple Syrup and Ginger Ale

Here is another recipe used as a replacement for canned or bottled soda. First we make a ginger flavored simple syrup, and use that to make ginger ale. This simple syrup will take a little longer as we want to develop a stronger ginger taste in the syrup since it is the only time this flavor will be added to the end product. The process is the same as above, with just a little time added.

Ingredients:

- 1 cup water
- 3/4 cup sugar
- 1 1/2 ounces grated ginger
- 2 tablespoons lemon juice
- 1 tablespoon lime juice
- Soda water

Time: 1 hour 15 minutes   Portions: 8   Serving size: 8 oz.

Place the sugar, water, ginger, and lemon juice in a pot and bring to a boil. Turn off heat and let steep for 45 minutes to an hour.

Strain the mixture to remove the ginger pieces and place in refrigerator.

When cool, add 2 tablespoon of ginger simple syrup to 8 ounces of soda water and 1 tablespoon of lime juice, serve over ice.

Ginger Simple Syrup

		Fat (g)	Sat. Fat (g)	Phos (mg)	Pot (mg)	Sodium (mg)	Calorie	Carb (g)	Vit A (iu)	Vit C (mg)	Vit K (Mcg)	Vit E (mg) AT	Vit B6 (mg)
	RDA	**65**	**20**	**700**	**3300**	**2300**	**2000**	**275**	**5000**	**60**	**120**	**20**	**2**
Ingredients	**Amt**												
Water	**1 cup**	0.00	0.00	0.00	0.00	0.00	0.00	0.00	0.00	0.00	0.00	0.00	0.00
		0.00%	0.00%	0.00%	0.00%	0.00%	0.00%	0.00%	0.00%	0.00%	0.00%	0.00%	0.00%
Sugar	**3/4 cup**	0	0	0	0	0	580.5	150	0	0	0	0	0
		0.00%	0.00%	0.00%	0.00%	0.00%	29.03%	54.55%	0.00%	64.80%	0.00%	0.00%	0.00%
Lemon	**1 oz**	0.00	0.00	13.80	36.10	1.50	13.00	3.00	0.80	2.80	0.20	0.00	0.10
		0.00%	0.00%	3.99%	1.09%	0.60%	0.65%	1.09%	0.02%	4.67%	0.17%	0.00%	5.00%
Grated fresh Ginger	**1 oz**	0.20	0.10	9.50	116.00	3.60	22.40	5.00	0.00	1.40	0.00	0.10	0.00
		0.31%	0.50%	1.36%	3.52%	0.16%	1.12%	1.82%	0.00%	2.33%	0.00%	0.50%	0.00%
Total for Dish		0.20	0.10	23.30	152.10	5.10	615.90	158.00	0.80	4.20	0.20	0.10	0.10
		0.31%	0.50%	3.33%	4.61%	0.22%	30.80%	57.45%	0.02%	7.00%	0.17%	0.50%	5.00%
# of Servings		6	6	6	6	6	6	6	6	6	6	6	4
Totals per serving		0.03	0.02	3.88	25.35	0.85	102.65	26.33	0.13	0.70	0.03	0.02	0.03
		0.05%	0.08%	0.55%	0.77%	0.04%	5.13%	9.58%	0.00%	1.17%	0.03%	0.08%	1.25%

Essential Amino Acid Chart for recipe Ginger Simple Syrup

	Mg/ gram protein needed	Sugar	Recipe 0 g	Lemon Juice/ Zest	Recipe .1 g	Giner Root, grated	Recipe .5 g	Total Gram	Total Amino Acid	Total Protein/g
Essential Amino Acids		Protein/g	Recipe	Protein/g	Recipe	Protein/g	Recipe			
Histidine	18	0.00	0.00	6.00	0.60	16.80	8.40	0.60	9.00	15.00
Isoleuciine	25	0.00	0.00	6.00	0.60	28.60	14.30		14.90	24.83
Methionine	25	0.00	0.00	6.00	0.60	7.20	3.60		4.20	7.00
Leucine	55	0.00	0.00	45.00	4.50	41.40	20.70		25.20	42.00
Lysine	51	0.00	0.00	45.00	4.50	32.00	16.00		20.50	34.17
Phenylalanine	47	0.00	0.00	31.00	3.10	25.20	12.60		15.70	26.17
Threonine	27	0.00	0.00	6.00	0.60	20.20	10.10		10.70	17.83
Tryptophan	7	0.00	0.00	6.00	0.60	3.40	1.70		2.30	3.83
Valine	32	0.00	0.00	31.00	3.10	40.80	20.40		23.50	39.17
Recipe Amount		1 Cup		2 Tbsp		2 Tbsp				
Protein per recipe (g)		0		0.1		0.5				

Total recipe Protein (g)	# of servings	Protein / serving (g)	Not a Complete Protein
0.60	6.00	0.10	

Building your own Tacos

Tacos have become a staple food all across the country, and for good reason. What some of us might remember as a hard crunchy shell, filled with ground beef cooked with a mysterious taco seasoning, yellow cheese, lettuce, and tomato, is now a food group all its own.

Hard or soft shell, flour or corn, flavored or traditional, are just the choices for the tortilla. The different combinations of protein, vegetables, cheeses, and sauces make the varieties seem endless.

The next several recipes are going to outline the nutritional issues of the pieces that make up tacos and how we can avoid some of the problem areas. I am going to address the building blocks individually and offer some alternatives.

The first step in taco-making is choosing the tortillas. The tortilla is the "bread" of this sandwich-style meal and it is the first thing to hit your tastebuds when you eat a taco (or sandwich).

Let's start with store-bought. Both the soft corn and flour style tortillas found in the grocery store are going to have at least 345 mg of sodium per tortilla. This works out to be about 15% of the 2300 mg RDI in the U.S. For those in other countries or on a restricted sodium diet of 1500 mg, this works out to be 23%. That is a lot of sodium just for the part of the taco that holds all the good stuff. This is in line with the levels of sodium in 2 slices of standard soft white bread.

Sodium offers flavor and just as important, shelf stability. Salt is a preservative and this is one reason tacos can be stored at room temperature on a store shelf for long periods of time. Other preservatives are also responsible for this wonder of what is now called food, and many are items CKD patients are told to avoid or limit.

Making your own may seem like a daunting task, but it is really rather simple. And after all, this is a cookbook, not a "Save time and effort by eating as many preservatives as possible" book.

Food Science Note

Many of us grew up being taught about the different parts of the tongue that are responsible for different tastes. Sweet, sour, salty, bitter, and sometimes umami. New evidence has shown that this is not the case and that all parts of the tongue can accept and taste all four (or five) flavors.

In addition, we have learned that sodium has the ability to open up more taste buds than other flavors. If a tastebud is opened up by sodium, it increases the ability of the same taste bud to accept other flavors. This means that more taste buds will accept sweet, sour, and bitter if first opened up by sodium. Think about all the salted sweet products that are now on the market. Salted caramel, chocolate, and others are popular, in part due to the increased taste sensation.

Not only does salt have its own flavor, it enhances the ability of other food items by allowing more taste buds to be receptive to flavors that normally would stay closed.

You can see why having a salty exterior of any food that hits the tongue first will enhance the overall flavor. Salt on top of our food, salty bread in a sandwich, or salty taco shell will all have the same effect.

Corn Tortillas

Corn Tortillas are made with only three ingredients. Masa harina, water, and salt. Masa harina is a corn flour that is very distinct both in flavor and how it is made. It is easily found in most large grocery stores or ones that have an ethnic section. It looks like a typical 5 pound bag of flour as is stored in the same section.

The recipe is quite simple. You will need something to flatten out the small balls of dough, and a heavy bottom pan or skillet to quickly fry the rounds into tortillas.

Ingredients:

- 1 1/2 cups masa harina
- 1/2 teaspoon salt
- 1 cup water, plus extra if needed.

Time: 45 minutes Yield: 9 tortillas Potions: 9
Serving size: 1 tortillas

Food Note

The "special process" mentioned above for making the corn flour is called nixtamalization. Corn kernels have tough outer layers that do not break down in the digestive system. This process removes much of the hard outer shell of the corn kernels allowing access to the nutrients inside. This increases bioavailability of all those hidden nutrients. The process cooks the kernels in lyme water and strips off the outer shell making a softer and more appealing dough.

It is believed that when the Spanish explorers brought corn to Europe from what is now central America, they did not bring the nixtamalization process. The effects of eating high levels of corn that was not processed caused many people to become very ill as their bodies were flushing out all the nutrients before being absorbed.

In a bowl large enough to hold all the ingredients, mix the masa harina and the salt. Add the water to make a ball of dough that will stick together, but will not stick to your hands. You may need to add a little water to get the dough to come together, or a bit more flour to make it less sticky. Humidity and how well packed the flour is when measured will play a role in this dough making process, so do not get discouraged if you need to alter the recipe a bit.

Wrap up the dough ball tightly in plastic and let rest for 30 minutes.

The next step is to heat your heavy bottom skillet to a medium high temperature. Cast iron works best for tortillas, but you can make do if you do not have one.

While the skillet is warming, divide the dough into 9 equal pieces, about 1.5 oz. each. Roll these into balls to assist with flattening into rounds.

There are several options to flatten out dough into rounds. These include the wonderful little tortilla makers, a rolling pin, or a heavy flat piece of, well, anything. I have small cutting boards that work, different small cast iron pans, and little pizza stones, all of which I have employed for the process.

I like to use parchment paper to place the dough between, but plastic wrap or a split plastic storage bag will also work. This prevents the dough from sticking to whatever you are using to flatten the tortilla. Be sure there are two separate pieces and not just one folded in half. Try to make them as thin as possible, and around 6" in diameter.

When you have created your rounds, place on the skillet for 1 to 2 minutes per side. When cooked, place on a towel and cover to keep warm. Alternatively you can keep the batch warm in an oven set on the lowest setting while you continue to cook your tortillas.

That is it. They're done. You can and should use them while warm, but they can be stored in plastic bags in the fridge, or even frozen. Re-warm them in the oven on low heat if not eaten right away. Perhaps the better option for leftovers is to cut them into triangles and fry for homemade tortillas chips. That is where mine usually end up.

Corn Tortilla

		Fat (g)	Sat. Fat (g)	Phos (mg)	Pot (mg)	Sodium (mg)	Calories	Carbs (g)	Vit A (iu)	Vit C (mg)	Vit K (Mcg)	Vit E (mg) AT	Vit B6 (mg)
	RDA	65	20	700	3300	2300	2000	275	5000	60	120	20	2
Ingredients	Amt												
Masa Harina	1 1/2 cup	6.00	1.50	381.00	510.00	8.70	624.00	131.00	366.00	0.00	0.00	0.00	0.60
		9.23%	7.50%	54.43%	15.45%	0.38%	31.20%	47.64%	7.32%	0.00%	0.00%	0.00%	30.00%
Kosher Salt	1/2 tsp					1150.00							
						50.00%							
Total for Dish		6.00	1.50	381.00	510.00	1158.70	624.00	131.00	366.00	0.00	0.00	0.00	0.60
		9.23%	7.50%	54.43%	15.45%	50.38%	31.20%	47.64%	7.32%	0.00%	0.00%	0.00%	30.00%
# of Servings		9	9	9	9	9	9	9	9	9	9	9	9
Totals per serving		0.67	0.17	42.33	56.67	128.74	69.33	14.56	40.67	0.00	0.00	0.00	0.07
		1.03%	0.83%	6.05%	1.72%	5.60%	3.47%	5.29%	0.81%	0.00%	0.00%	0.00%	3.33%

Essential Amino Acid Chart for recipe Corn Tortilla

	Mg/ gram protein needed	Masa Harina	Recipe 15.9 g			Total Gram s	Total Amino Acid	Total Protein /g
Essential Amino Acids		**Protein/g**	**Recipe**					
Histidine	**18**	30.66	487.50			15.90	487.50	30.66
Isoleuciine	**25**	35.94	571.50				571.50	35.94
Methionine	**25**	21.04	334.50				334.50	21.04
Leucine	**55**	123.21	1959.00				1959.00	123.21
Lysine	**51**	28.30	450.00				450.00	28.30
Phenylalanin e	**47**	49.34	784.50				784.50	49.34
Threonine	**27**	37.74	600.00				600.00	37.74
Tryptophan	**7**	7.09	112.80				112.80	7.09
Valine	**32**	50.85	808.50				808.50	50.85
Recipe Amount		1 1/2 cup						
Protein per recipe (g)		15.90						

Total recipe Protein (g)	# of servings	Protein / serving (g)	Not a Complete Protein
15.90	9.00	1.77	

Flour Tortillas

Flour tortillas are very similar to corn. There is one ingredient replacement and one added, but it is not necessary. The process is also very similar with the exception that I prefer to use a rolling pin or a small dowel from the hardware store to form the discs.

Flour and corn tortillas have a very different taste and the choice is up to the eater. Flour tortillas, however, are lower in phosphorus so long as you use white flour only. This is one of the many deviations in the CKD diet from standard programs we think of as healthy eating. Whole wheat, rye, and other grain flours have higher levels of phosphorus, and it is currently suggested that CKD patients stick to white flour.

Health Note

Phosphorus levels are one of the problems many CKD patients battle. Remember that if the kidneys are not functioning properly, waste is not being removed at the proper rate, or at all. Our bodies need phosphorus and it is found in nearly all the food we eat. That is good news for kidneys that are keeping things in balance. Not so good for those with limited function.The excess can build up in veins, arteries, and around our organs.

Noori, et. al. (2010) determined that there is an absorption difference between the phosphorus found naturally in food (organic phosphorus) and the types that are added to food through the use of chemicals (inorganic phosphorus). Organic phosphorus absorbs into the body at a much lower level than inorganic phosphorus. This means that the processed foods and shelf stable products that have added phosphorus are much more dangerous for CKD patients.

The phosphorus levels listed in the recipes of this book are organic levels as I do not use any shelf stabilizing products or added phosphates. I know that this is not always an option, so when varying the recipe, please read those labels carefully.

The sodium levels in store bought flour tortillas are the same as store bought corn. You are reducing both sodium and phosphorus levels by making your own flour tortillas. Many store bought products will have added phosphorus to help prolong shelf life and appearance. Phosphorus is not listed in the nutrition information, but you can often find it listed somewhere in the ingredients.

Ingredients:

- 2 cups all purpose flour
- 3 tablespoons lard or vegetable oil (optional - if fat or saturated fat are an issue, eliminate)
- 1/4 teaspoon salt
- 1/2 cup water (more if necessary)

Time: 45 minutes Yield: 9 tortillas Potions: 9
Serving size: 1 tortillas

Mix the flour and salt together, add the lard and crumble between your fingers while in the flour. Next add the water and mix creating a dry dough that just comes together. You can use a mixer, food processor, or mix by hand.

Wrap the dough tightly in plastic and let rest for 1 hour. Divide the dough into 9 portions and roll each into a round disk using a rolling pin.

Using a cast iron skillet or heavy bottom pan, cook each round on medium high heat for 1 to 2 minutes on each side. You should notice some browning in areas on the heat side and this is a good indicator the tortilla is ready to flip or is done.

Place on a towel or keep warm in an oven until ready to eat.

Food Note

At this point many of you may be wondering why would lard be acceptable for CKD patients, ever. There are two reasons I will give you. The first is called "mouth feel". A good flour tortilla creates a feel on the tongue and also adds flavor to a potentially bland item.

The second point is the fat and phosphorus levels. You can look at the phosphorus/protein chart in the back section of this book for the details. You will notice that lard has more total fat, but less saturated fat than butter. The fat levels in margarine are less, provided it is not partially hydrogenated, but both butter and margarine have more phosphorus. Lard has no phosphorus at all.

Depending on what particular issue is more important for your diet (check with your dietitian, please) lard may be a better option than butter or margarine.

Flour Tortilla-1

		Fat (g)	Sat. Fat (g)	Phos (mg)	Pot (mg)	Sodium (mg)	Calories	Carb (g)	Vit A (iu)	Vit C (mg)	Vit K (Mcg)	Vit E (mg) AT	Vit B6 (mg)
	RDA	**65**	**20**	**700**	**3300**	**2300**	**2000**	**275**	**5000**	**60**	**120**	**20**	**2**
Ingredients	**Amt**												
All Purpose Flour		2.40	0.40	270.00	268.00	5.00	910.00	131.00	5.00	0.00	0.80	0.60	0.20
		3.69%	2.00%	38.57%	8.12%	0.22%	45.50%	47.64%	0.10%	0.00%	0.67%	3.00%	10.00%
Lard	**3 Tbsp**	38.40	15.00	0.00	0.00	0.00	445.00	0.00	0.00	0.00	0.00	0.00	0.00
		59.08%	75.00%	0.00%	0.00%	0.00%	22.25%	0.00%	0.00%	0.00%	0.00%	0.00%	0.00%
Kosher Salt	**1/2 tsp**					1150.00							
						50.00%							
Total for Dish		40.80	15.40	270.00	268.00	1155.00	1355.00	131.00	5.00	0.00	0.80	0.60	0.20
		62.77%	77.00%	38.57%	8.12%	50.22%	67.75%	47.64%	0.10%	0.00%	0.67%	3.00%	10.00%
# of Servings		8	8	8	8	8	8	8	8	8	8	8	8
(1 Tbsp)													
Totals per serving		5.10	1.93	33.75	33.50	144.38	169.38	16.38	0.63	0.00	0.10	0.08	0.03
		7.85%	9.63%	4.82%	1.02%	6.28%	8.47%	5.95%	0.01%	0.00%	0.08%	0.38%	1.25%

Essential Amino Acid Chart for Flour Tortilla

	Mg/ gram protein needed	Flour	Recipe 25.8 g	Lard	Recipe 0 g			Total Grams	Total Amino Acid	Total Protein/g
Essential Amino Acids		Protein/g	Recipe	Protein/g	Recipe					
Histidine	18	22.33	576.00	0.00	0.00			25.80	576.00	22.33
Isoleuciine	25	34.57	892.00	0.00	0.00				892.00	34.57
Methionine	25	17.75	458.00	0.00	0.00				458.00	17.75
Leucine	55	68.84	1776.00	0.00	0.00				1776.00	68.84
Lysine	51	22.09	570.00	0.00	0.00				570.00	22.09
Phenylalanine	47	50.39	1300.00	0.00	0.00				1300.00	50.39
Threonine	27	27.21	702.00	0.00	0.00				702.00	27.21
Tryptophan	7	12.33	318.00	0.00	0.00				318.00	12.33
Valine	32	40.23	1038.00	0.00	0.00				1038.00	40.23
Recipe Amount		2 cups		3Tbsp						
Protein per recipe (g)	8.00	25.80		0.00						

Total recipe Protein (g)	# of servings	Protein / serving (g)	Not a Complete Protein
25.80	9.00	2.87	

Flavored tortillas and leftovers

One of the ways these recipes are better for CKD patients, or frankly anyone, is the significantly lower level of sodium. Each recipe has less than half of the sodium for each serving. It can be higher given that corn tortillas are often doubled up because of potential cracks in the dough that can cause leaking.

One of the problems with the lower sodium levels is the diminished taste. Although the tortillas are only the wrapper for all the food and flavors inside, it is still the first flavor to hit the tongue.

In the event the lower levels of sodium are not appealing enough, we could easily add some ingredients to enhance the flavor. This is one of the ideas that many nutritionists suggest for other foods as they try to assist CKD patients with menu items.

The process is the same as above with the addition of a few ingredients. The recipes below will provide the ingredients and the nutritional information, but not the process.

Chipotle Corn Tortilla

Ingredients:

- 1 1/2 cups masa harina
- 2 chipotle peppers with adobo sauce
- 1 cup water

First, combine the water, peppers, sauce and mix well together. This can be done in a food processor or in a tall jar using an immersion blender. Once this is done, use this liquid as the water in the previous recipes above.

I have left out the salt in this recipe as the peppers and adobo sauce have enough sodium.

Chipotle Corn Tortilla

		Fat (g)	Sat. Fat (g)	Phos (mg)	Pot (mg)	Sodium (mg)	Calories	Carb (g)	Vit A (iu)	Vit C (mg)	Vit K (Mcg)	Vit E (mg) AT	Vit B6 (mg)
	RDA	**65**	**20**	**700**	**3300**	**2300**	**2000**	**275**	**5000**	**60**	**120**	**20**	**2**
Ingredients	**Amt**												
Masa Harina,	**1 1/2 cup**	6.00	1.50	381.00	510.00	8.70	624.00	131.00	366.00	0.00	0.00	0.00	0.60
		9.23%	7.50%	54.43%	15.45%	0.38%	31.20%	47.64%	7.32%	0.00%	0.00%	0.00%	30.00%
Chipotle Peppers and adobo sauce	**2 plus sauce**	0.00	0.00	8.00	85.00	736.00	12.00	2.00	728.00	4.40	5.60	0.40	0.00
		0.00%	0.00%	1.14%	2.58%	32.00%	0.60%	0.73%	14.56%	64.80%	4.67%	2.00%	0.00%
Total for Dish		6.00	1.50	389.00	595.00	744.70	636.00	133.00	1094.00	4.40	5.60	0.40	0.60
		9.23%	7.50%	55.57%	18.03%	32.38%	31.80%	48.36%	21.88%	7.33%	4.67%	2.00%	30.00%
# of Servings		9	9	9	9	9	9	9	9	9	9	9	9
Totals per serving		0.67	0.17	43.22	66.11	82.74	70.67	14.78	121.56	0.49	0.62	0.04	0.07
		1.03%	0.83%	6.17%	2.00%	3.60%	3.53%	5.37%	2.43%	0.81%	0.52%	0.22%	3.33%

Essential Amino Acid Chart for recipe Chipotle Corn Tortilla

	Mg/ gram protein needed	Masa Harina	Recipe 15.9 g	Chipotle Peppers with adobo	Recipe .4 g			Total Grams	Total Amino Acid	Total Protein/g
Essential Amino Acids		Protein/g	Recipe	Protein/g	Recipe					
Histidine	18	30.66	487.50	20.00	8.00			16.30	495.50	30.40
Isoleuciine	25	35.94	571.50	33.00	13.20				584.70	35.87
Methionine	25	21.04	334.50	13.00	5.20				339.70	20.84
Leucine	55	123.21	1959.00	53.00	21.20				1980.20	121.48
Lysine	51	28.30	450.00	46.00	18.40				468.40	28.74
Phenylalanine	47	49.34	784.50	32.00	12.80				797.30	48.91
Threonine	27	37.74	600.00	36.50	14.60				614.60	37.71
Tryptophan	7	7.09	112.80	13.00	5.20				118.00	7.24
Valine	32	50.85	808.50	43.00	17.20				825.70	50.66
Recipe Amount		1 1/2 cup		2 peppers						
Protein per recipe (g)	1.81	15.90		0.40						

Total recipe Protein (g)	# of servings	Protein / serving (g)	Not a Complete Protein
16.30	9.00	1.81	

Lime Cilantro Corn Tortilla

Ingredients:

- 1 1/2 cups masa harina
- 1 tablespoon fresh cilantro
- 1 lime, juice and zest
- 1/4 teaspoon kosher salt
- 1 cup water

Combine cilantro, lime juice, zest, and salt with the water. Mix thoroughly in a food processor or with an immersion blender. Continue as above with the masa harina, forming and cooking the tortillas.

Any of the corn tortillas can be used to make tortilla chips, baked or fried. Simply cut into triangles and place evenly on a cooling rack fitted into a baking sheet. Back at 375° F for 15 to 20 minutes, depending on how thick your tortillas are.

To fry, heat canola or peanut oil in a heavy bottom pot and fry at 375° F for 1 minute on each side.

Add flavoring or a small amount of salt when hot, and serve.

Lime Cilantro CornTortilla

		Fat (g)	Sat. Fat (g)	Phos (mg)	Pot (mg)	Sodium (mg)	Calories	Carb (g)	Vit A (iu)	Vit C (mg)	Vit K (Mcg)	Vit E (mg) AT	Vit B6 (mg)
	RDA	**65**	**20**	**700**	**3300**	**2300**	**2000**	**275**	**5000**	**60**	**120**	**20**	**2**
Ingredients	**Amt**												
Masa Harina	**1 1/2 cup**	6.00	1.50	381.00	510.00	8.70	624.00	131.00	366.00	0.00	0.00	0.00	0.60
		9.23%	7.50%	54.43%	15.45%	0.38%	31.20%	47.64%	7.32%	0.00%	0.00%	0.00%	30.00%
Kosher Salt	**1/4 tsp**	0.00	0.00	0.00	0.00	575.00	0.00	0.00	0.00	0.00	0.00	0.00	0.00
		0.00%	0.00%	0.00%	0.00%	25.00%	0.00%	0.00%	0.00%	64.80%	0.00%	0.00%	0.00%
Cilantro, fresh	**1 tsp**	0.80	0.00	53.60	584.00	51.60	25.60	4.00	7556.00	30.40	347.20	2.80	0.00
		1.23%	0.00%	7.66%	17.70%	2.24%	1.28%	1.45%	151.12%	50.67%	289.33%	14.00%	0.00%
Lime juice and zest	**1 Tbsp**	0.00	0.00	3.99	32.80	0.50	7.00	2.00	14.00	8.40	0.20	0.10	0.00
Total for Dish		6.80	1.50	438.59	1126.80	635.80	656.60	137.00	7936.00	38.80	347.40	2.90	0.60
		10.46%	7.50%	62.66%	34.15%	27.64%	32.83%	49.82%	158.72%	64.67%	289.50%	14.50%	30.00%
# of Servings		9	9	9	9	9	9	9	9	9	9	9	9
(1 Tbsp)													
Totals per serving		0.76	0.17	48.73	125.20	70.64	72.96	15.22	881.78	4.31	38.60	0.32	0.07
		1.16%	0.83%	6.96%	3.79%	3.07%	3.65%	5.54%	17.64%	7.19%	32.17%	1.61%	3.33%

Essential Amino Acid Chart for Lime Cilantro Tortilla

	Mg/ gram protein needed	Masa Harina	Recipe 15.9 g			Lime juice	Recipe .5 g	Total Grams	Total Amino Acid	Total Protein/ g
Essential Amino Acids		**Protein/g**	**Recipe**	**Protein/g**	**Recipe**	**Protein/ g**				
Histidine	**18**	30.66	487.50			6.00	3.00	16.40	487.50	29.73
Isoleuciine	**25**	35.94	571.50			6.00	3.00		571.50	34.85
Methionine	**25**	21.04	334.50			6.00	3.00		334.50	20.40
Leucine	**55**	123.21	1959.00			45.00	22.50		1959.00	119.45
Lysine	**51**	28.30	450.00			45.00	22.50		450.00	27.44
Phenylalanine	**47**	49.34	784.50			31.00	15.50		784.50	47.84
Threonine	**27**	37.74	600.00			6.00	3.00		600.00	36.59
Tryptophan	**7**	7.09	112.80			6.00	3.00		112.80	6.88
Valine	**32**	50.85	808.50			31.00	15.50		808.50	49.30
Recipe Amount		1 1/2 cup				1 Tbsp				
Protein per recipe (g)		15.90				0.50				

Total recipe Protein (g)	**# of servings**	**Protein / serving (g)**	Not a Complete Protein
16.40	9.00	1.82	

Salsas

Salsa is a general term that means "sauce" in Spanish. In food use it can apply to anything from a cooked red tomato and pepper dip, that looks like a typical sauce, to a colorful blend of raw veggies and fruits.

It is served as a dip for just about anything and it is also considered a primary add-on when making tacos.

Because tomatoes are high in potassium, many CKD patients are told to limit them in their diet. When using a tomato based salsa as a dip, this warning is well advised based on the amounts that you may consume.

However, when using as an add-on to tacos, the amounts of salsa, and hence the potassium are much lower. I have used a 2 tablespoon amount per taco in the following salsas, which I believe to be a bit more that most people might use.

The uncooked chunky version of salsa is called Pico de Gallo (translation: Beak of the Rooster) or Salsa Fresco (fresh salsa), and this is the type of tomato salsa I use for the recipes. If you use this recipe as a type of dip for tortillas, you will most likely consume a much higher level of potassium, and it may not be recommended.

As always, please consult your health care team about your specific needs and even present the data and charts for a better coordinated effort in managing your health and diet.

Pico de Gallo

Ingredients:

- 4 plum tomatoes, diced
- 1/4 cup shallot (1 large), minced
- 1 lime, juice and zest
- 2 serrano or jalapeño peppers, minced
- 1/4 cup green onion, minced
- 2 tablespoons cilantro, chopped
- 1/2 teaspoon salt

Total Time: 15 minutes Yield: 1 1/2 cups Portions: 12
Serving size : 2 Tbps.

Combine all the ingredients into a bowl and mix well. It's ready to use right away or can be stored in the fridge for up to 2 days. Note that the cold temperatures will reduce the flavor of the tomatoes, the salt will remove the water from the veggies, and the mixture will get soupy over time.

Pico de Gallo

		Fat (g)	Sat. Fat (g)	Phos (mg)	Pot (mg)	Sodium (mg)	Calories	Carb (g)	Vit A (iu)	Vit C (mg)	Vit K (Mcg)	Vit E (mg) AT	Vit B6 (mg)
	RDA	65	20	700	3300	2300	2000	275	5000	60	120	20	2
Ingredients	Amt												
Plum Tomato	4	0.40	0.00	59.60	588.00	12.40	44.80	9.60	2064.00	31.60	19.60	1.20	0.00
248 g/ 9 oz		0.62%	0.00%	8.51%	17.82%	0.54%	2.24%	3.49%	41.28%	52.67%	16.33%	6.00%	0.00%
Shallot, diced	1/4 cup	0	0	24	133.6	4.8	28.8	8	476	3.2	0	0	0
1 large (17 g)		0.00%	0.00%	3.43%	4.05%	0.21%	1.44%	2.91%	9.52%	64.80%	0.00%	0.00%	0.00%
Lime Juice and zest-fresh	1 oz	0.00	0.00	8.00	34.70	1.20	14.00	3.00	14.00	8.40	0.20	0.10	0.00
(2 Tbsp)		0.00%	0.00%	3.99%	1.05%	0.60%	0.70%	1.09%	0.28%	14.00%	0.17%	0.50%	0.00%
Jalapeño or Serrano pepper	2	0.00	0.00	8.60	62.20	0.20	8.00	2.00	224.00	12.40	2.80	0.20	0.20
		0.00%	0.00%	1.23%	1.88%	0.01%	0.40%	0.73%	4.48%	20.67%	2.33%	1.00%	10.00%
Green onion	1/4 cup	0.00	0.00	9.25	69.00	4.00	4.00	1.80	249.25	4.70	51.75	0.13	0.03
		0.00%	0.00%	1.32%	2.09%	0.17%	0.20%	0.65%	4.99%	7.83%	43.13%	0.63%	1.25%
Cilantro, fresh	2 Tbsp	0.10	0.00	13.40	146.00	12.90	6.40	1.00	1889.00	7.60	86.80	0.70	0.00
		0.15%	0.00%	1.91%	4.42%	0.56%	0.32%	0.36%	37.78%	12.67%	72.33%	3.50%	0.00%
Salt		0.00		0.00	0.00	575.00							
		0.00%	0.00%	0.00%	0.00%	25.00%	0.00%	0.00%	0.00%	0.00%	0.00%	0.00%	0.00%
Total for Dish	1 1/2 cups	0.50	0.00	122.85	1033.50	610.50	106.00	25.40	4916.25	67.90	161.15	2.33	0.23
		0.77%	0.00%	17.55%	31.32%	26.54%	5.30%	9.24%	98.33%	113.17%	134.29%	11.63%	11.25%
# Servings		12	12	12	12	12	12	12	12	12	12	12	12
Totals per serving		0.04	0.00	10.24	86.13	50.88	8.83	2.12	409.69	5.66	13.43	0.19	0.02
	8 6/7	0.06%	0.00%	1.46%	2.61%	2.21%	0.44%	0.77%	8.19%	9.43%	11.19%	0.97%	0.94%

Essential Amino Acid Chart for Pico de Gallo

	Mg/ gram protein needed	Tomato, plum	Recipe 2 g	Shallot	Recipe 1.2 g	Lime Juice, fresh	Recipe .1 g	Green Onion	Recipe .45 g	Total Gram s	Total Amin o Acid	Total Protein/g
Essential Amino Acids		Protein/g	Recipe	Protein/g	Recipe	Protein/g	Recipe	Protei n/g	Recipe			
Tryptophan	7	7.40	14.80	9.33	11.20	13.50	1.35	11.11	5.00	3.75	32.35	8.63
Histidine	18	17.40	34.80	14.33	17.20	38.70	3.87	17.78	0.00		55.87	14.90
Isoleuciine	25	22.40	44.80	35.33	42.40	53.00	5.30	42.78	0.00		92.50	24.67
Methionine	25	74.00	148.00	9.00	10.80	26.80	2.68	11.11	0.00		161.48	43.06
Threonine	27	33.40	66.80	32.67	39.20	36.90	3.69	40.00	0.00		109.69	29.25
Valine	32	22.40	44.80	36.67	44.00	68.70	6.87	45.00	0.00		95.67	25.51
Phenylalanine	47	83.00	166.00	27.00	32.40	53.80	5.38	32.78	0.00		203.78	54.34
Lysine	51	33.40	66.80	41.67	50.00	92.58	9.26	50.56	0.00		126.06	33.62
Leucine	55	31.00	62.00	49.67	59.60	96.70	9.67	60.56	0.00		131.27	35.01
Recipe Amount		4 each		1/4 cup		1 each		1/4 cup				
Protein per recipe (g)		2		1.2		0.1		0.45				

Total recipe Protein (g)	# of servings	Protein / serving (g)	Not a Complete Protein
3.75	12.00	0.31	

Fresh Mango Salsa

This is a very colorful dish, and because mangos have less potassium than tomatoes, it is a good substitute for your tacos. It goes especially well with chicken, pork, and fish.

It is NOT low enough in potassium to be eaten as a dip with tortillas, unless you are willing to limit the portion size to under a cup. This will still give you about 190 mg of potassium (about 12% of your RDI).

The recipe is simple as long as you can master one of the techniques for getting at the mango "meat."

Ingredients:

- 2 mangoes, peeled and cubed (see below)
- 1 cup red pepper, diced
- 2 jalapeño or serrano peppers, diced
- 1/4 cup green onion, chopped
- 2 tablespoons fresh cilantro, chopped
- 1 lime, juice and zest
- 1/2 teaspoon salt

Total time: 15 minutes Yield: 3 cups Portions: 24

Serving size: 2 Tbsps.

Gently mix all the ingredients together. Ripe mangoes are very soft and will break apart if mixed too vigorously.

Peeling a mango: Unlike an avocado, the pit of a mango does not peel away from the "meat" easily, if at all. The skin is also a bit more clingy and it can be difficult to manage.

The pit is not round, but shaped like a large almond. Check the bottom of the mango to help determine which is the wide side of the pit and which in narrow. When cutting the fruit off the pit, slice so that the blade of the knife and the wide part of the pit run parallel. This will help get the most yield out of each mango.

There are a few ways to peel a mango, and I will go over two of those methods here.

One: Cut each side of the mango off leaving the skin on. Be sure to cut the fruit as listed above. This will leave you with two halves with the skin.

With the skin side down, score each half with parallel cuts 1/2" apart, as deep as you can without cutting through the skin. Repeat with cuts perpendicular to the first, creating little squares. Push the bottom of the skin side from the middle up to reverse the arc of the mango. Now run a knife between the "meat" and the skin to remove the little squares.

Two: Peel the skin away from the mango using a peeler or knife. A peeler may may be more difficult, but it can be done. Be sure to cut a small piece off the top and bottom to create a more stable surface. Start from the top and run your knife below the skin and down to the bottom, arcing the cut with the curve of the mango. Repeat all the way around. This will leave you with a very slippery fruit to handle, so be careful. Next, cut the mango halves as above, and then into cubes.

Mango Salsa

		Fat (g)	Sat. Fat (g)	Phos (mg)	Pot (mg)	Sodium (mg)	Calories	Carb (g)	Vit A (iu)	Vit C (mg)	Vit K (Mcg)	Vit E (mg) AT	Vit B6 (mg)
	RDA	**65**	**20**	**700**	**3300**	**2300**	**2000**	**275**	**5000**	**60**	**120**	**20**	**2**
Ingredients	**Amt**												
Mango, Fresh	**2 cups**	0.80	0.20	36.40	514.00	6.60	214.00	56.20	2524.00	91.40	13.80	3.60	0.40
(2 Mangos)		1.23%	1.00%	5.20%	15.58%	0.29%	10.70%	20.44%	50.48%	152.33	11.50%	18.00%	20.00%
Red Pepper	**1 cup**	0.4	0	38.6	314	6	46.2	99	4666	190	7.3	2.4	0.4
		0.62%	0.00%	5.51%	9.52%	0.26%	2.31%	36.00%	93.32%	64.80	6.08%	12.00%	20.00%
Lime Juice and zest-fresh	**1 oz**	0.00	0.00	8.00	34.70	1.20	14.00	3.00	14.00	8.40	0.20	0.10	0.00
(2 Tbsp)		0.00%	0.00%	3.99%	1.05%	0.60%	0.70%	1.09%	0.28%	14.00	0.17%	0.50%	0.00%
Jalapeño or Serrano pepper	**2**	0.00	0.00	8.60	62.20	0.20	8.00	2.00	224.00	12.40	2.80	0.20	0.20
		0.00%	0.00%	1.23%	1.88%	0.01%	0.40%	0.73%	4.48%	20.67	2.33%	1.00%	10.00%
Green onion	**1/4 cup**	0.00	0.00	9.25	69.00	4.00	4.00	1.80	249.25	4.70	51.75	0.13	0.03
		0.00%	0.00%	1.32%	2.09%	0.17%	0.20%	0.65%	4.99%	7.83%	43.13%	0.63%	1.25%
Cilantro, fresh	**2 Tbsp**	0.10	0.00	13.40	146.00	12.90	6.40	1.00	1889.00	7.60	86.80	0.70	0.00
		0.15%	0.00%	1.91%	4.42%	0.56%	0.32%	0.36%	37.78%	12.67	72.33%	3.50%	0.00%
Salt		0.00		0.00	0.00	575.00							
		0.00%	0.00%	0.00%	0.00%	25.00%	0.00%	0.00%	0.00%	0.00%	0.00%	0.00%	0.00%
Total for Dish	**3 cups**	1.30	0.20	114.25	1139.9	605.90	292.60	163.00	9566.25	314.50	162.65	7.13	1.03
		2.00%	1.00%	16.32%	34.54%	26.34%	14.63%	59.27%	191.33%	524.17	135.54%	35.63%	51.25%
2 Tbsp per		24	24	24	24	24	24	24	24	24	24	24	24
Totals per serving		0.05	0.01	4.76	47.50	25.25	12.19	6.79	398.59	13.10	6.78	0.30	0.04
		0.08%	0.04%	0.68%	1.44%	1.10%	0.61%	2.47%	7.97%	21.84	5.65%	1.48%	2.14%

Essential Amino Acid Chart for recipe Mango Salsa

	Mg/ gram protein needed	Mango, fresh	Recipe 1.6 g	Pepper, red, raw	Recipe 1.5 g	Lime Juice, fresh	Recipe .1 g	Green Onion	Recip e .45 g	Total Grams	Total Amino Acid	Total Protein/g
Essential Amino Acids		Protein/g	Recipe	Protein/g	Recipe	Protein/g	Recipe	Protein/g	Recip e			
Histidine	18	23.50	37.60	16.87	25.30	38.70	3.87	17.78	8.00	3.65	74.77	20.48
Isoleuciine	25	37.13	59.40	20.87	31.30	53.00	5.30	42.78	19.25		115.25	31.58
Methionine	25	10.38	16.60	5.93	8.90	26.80	2.68	11.11	5.00		33.18	9.09
Leucine	55	64.00	102.40	35.73	53.60	96.70	9.67	60.56	27.25		192.92	52.85
Lysine	51	84.50	135.20	35.73	53.60	92.58	9.26	50.56	22.75		220.81	60.50
Phenylalanine	47	35.13	56.20	49.67	74.50	53.80	5.38	32.78	14.75		150.83	41.32
Threonine	27	39.13	62.60	39.73	59.60	36.90	3.69	40.00	18.00		143.89	39.42
Tryptophan	7	16.50	26.40	11.93	17.90	13.50	1.35	11.11	5.00		50.65	13.88
Valine	32	53.63	85.80	30.80	46.20	68.70	6.87	45.00	20.25		159.12	43.59
Recipe Amout		2 cups		1 cup		1 each		1/4 cup				
Protein per recipe (g)		1.6		1.5		0.1		0.45				

Total recipe Protein (g)	# of servings	Protein / serving (g)	Not a Complete Protein
3.65	24.00	0.15	

Crema

Mexican crema is a mild sour cream that is used on a variety of dishes, similar to the French creme fraiche or sour cream.

It is quite easy to make at home and with a little extra effort you can add additional flavors to boost the taste.

Crema, creme fraiche, and sour cream are all made with heavy cream and are going to be high in fat. The amount used in taco recipes and on most other dishes, however, will limit the total fat in the final dish. I do not recommend that you eat this by the spoonful, especially if you are on any type of reduced fat diet.

There are two benefits to using heavy cream and they are the same points made in other recipes. The first is the reduction in phosphorus when it is used rather than other milk products. One cup of whole milk has 232 grams of phosphorus while 1 cup of heavy cream has about 148 grams. It should also be noted that in almost all cases you will never use the same amounts of milk and heavy cream in a recipe as the cream is very thick and much more flavorful.

The second benefit is the ability of fat to carry flavor to the whole dish. Fat absorbs flavors and the higher levels of fat in cream indicates you can bring a great deal more flavor to any dish, thus limiting the need to increase sodium levels to jump start the tastebuds.

Ingredients:

- 1 cup heavy whipping cream
- 1 tablespoon buttermilk

Total time: 24 hours   Yield: 1 cup   Portions: 16

Serving size: 1 Tbsp

Place the heavy cream and the buttermilk in a jar and loosely cover. Leave the jar on the counter at room temperature for 8 to 12 hours. The mixture will look a little lumpy when ready, and at this point mix it well. Cover tightly and place in the refrigerator. It is ready for use.

Creme Fraiche or Crema

		Fat (g)	Sat. Fat (g)	Phos (mg)	Pot (mg)	Sodium (mg)	Calories	Carb (g)	Vit A (iu)	Vit C (mg)	Vit K (Mcg)	Vit E (mg) AT	Vit B6 (mg)
	RDA	**65**	**20**	**700**	**3300**	**2300**	**2000**	**275**	**5000**	**60**	**120**	**20**	**2**
Ingredients	**Amt**												
Heavy Whipping Cream	**1 cup**	88.00	55.00	148.00	178.00	90.40	821.00	7.00	3499.00	1.40	7.60	2.50	0.10
		135.38%	275.00%	21.14%	5.39%	3.93%	41.05%	2.55%	69.98%	2.33%	6.33%	12.50%	5.00%
Buttermilk	**1 Tbsp**	0.14	0.08	13.63	23.13	16.06	6.13	0.75	3.98	0.00	0.01	0.01	0.01
		0.21%	0.41%	1.95%	0.70%	0.70%	0.31%	0.27%	0.08%	64.80%	0.01%	0.03%	0.31%
Total for Dish		88.14	55.08	161.63	201.13	106.46	827.13	7.75	3502.98	1.40	7.61	2.51	0.11
! cup + 1 Tbsp		135.60%	275.41%	23.09%	6.09%	4.63%	41.36%	2.82%	70.06%	2.33%	6.34%	12.53%	5.31%
# of Servings		16	16	16	16	16	16	16	16	16	16	16	16
(1 Tbsp)													
Totals per serving		5.51	3.44	10.10	12.57	6.65	51.70	0.48	218.94	0.09	0.48	0.16	0.01
		8.47%	17.21%	1.44%	0.38%	0.29%	2.58%	0.18%	4.38%	0.15%	0.40%	0.78%	0.33%

Essential Amino Acid Chart for recipe Crema

	Mg/ gram protein needed	Heavy Whipping Cream	Recipe 4.9 g	ButterMilk	Recipe .51 g			Total Grams	Total Amino Acid	Total Protein/g
Essential Amino Acids		Protein/g	Recipe	Protein/g	Recipe	Protein/g	Recipe			
Histidine	18	27.14	133.00	6.00	3.04			5.41	136.04	25.16
Isoleuciine	25	60.20	295.00	6.00	3.04				298.04	55.13
Methionine	25	24.69	121.00	6.00	3.04				124.04	22.94
Leucine	55	97.55	478.00	45.00	22.78				500.78	92.63
Lysine	51	79.18	388.00	45.00	22.78				410.78	75.98
Phenylalanine	47	48.16	236.00	31.00	15.69				251.69	46.56
Threonine	27	45.10	221.00	6.00	3.04				224.04	41.44
Tryptophan	7	14.08	69.00	6.00	3.04				72.04	13.32
Valine	32	66.53	326.00	31.00	15.69				341.69	63.20
Recipe Amount		1 Cup		1 Tbsp						
Protein per recipe (g)		4.90		0.51						

Total recipe Protein (g)	# of servings	Protein / serving (g)	Complete Protein
5.41	16.00	0.34	

Chipotle Crema

Using the same system as above, this recipe adds a smokey and spicy addition to the slightly sour cream taste of the base crema. It is important to point out that dairy will subdue the spiciness in food and if you want a spicier product you will need to add more of the hotter items in the recipe.

Ingredients:

- 1 cup heavy whipping cream
- 1 tablespoon buttermilk
- 2 chipotle peppers and adobo sauce

Total time: 24 hours   Yield: 1 cup   Portions: 16

Serving size: 1Tbsp

Make the crema as before and then add the chopped chipotle peppers and the adobo sauce to the mixture. Refrigerate when combined.

Food Note

The capsaicin in peppers binds to a receptor that exists in many parts of the body, including the tongue. When the tongue receptors come in contact with the capsaicin, the receptors send a signal to the brain and it returns a signal of heat or abrasion on the tongue.
According to the American Chemical society, the protein called casein found in milk and milk products, binds to the capsaicin and helps remove it from the receptors the tongue. Water does not have the same effect, and merely spreads the heat around the tongue.
Soy, almond, and other plant-based milky waters will not have the same soothing effect because they do not have any casein proteins.

Chipotle Crema

		Fat (g)	Sat. Fat (g)	Phos (mg)	Pot (mg)	Sodium (mg)	Calories	Carb (g)	Vit A (iu)	Vit C (mg)	Vit K (Mcg)	Vit E (mg) AT	Vit B6 (mg)
	RDA	**65**	**20**	**700**	**3300**	**2300**	**2000**	**275**	**5000**	**60**	**120**	**20**	**2**
Ingredients	**Amt**												
Creme Fraiche	**1 cup**	88.14	55.08	161.63	201.13	106.46	827.13	7.75	3502.9	1.40	7.61	2.51	0.11
		135.60%	275.41%	23.09%	6.09%	4.63%	41.36%	2.82%	70.06%	2.33%	6.34%	12.53%	5.31%
Chipotle Peppers and adobo sauce	**2 Tbsp**	0.00	0.00	8.00	85.00	736.00	12.00	2.00	728.00	4.40	5.60	0.40	0.00
		0.00%	0.00%	1.14%	2.58%	32.00%	0.60%	0.73%	14.56%	64.80%	4.67%	2.00%	0.00%
Total for Dish		88.14	55.08	169.63	286.13	842.46	839.13	9.75	4230.9	5.80	13.21	2.91	0.11
		135.60%	275.41%	24.23%	8.67%	36.63%	41.96%	3.55%	84.62%	9.67%	11.01%	14.53%	5.31%
# of Servings		16	16	16	16	16	16	16	16	16	16	16	16
(1 Tbsp)													
Totals per serving		5.51	3.44	10.60	17.88	52.65	52.45	0.61	264.44	0.36	0.83	0.18	0.01
		8.47%	17.21%	1.51%	0.54%	2.29%	2.62%	0.22%	5.29%	0.60%	0.69%	0.91%	0.33%

Essential Amino Acid Chart for Chipotle Crema

	Mg/ gram protein needed	Creme Fraiche	Recipe 5.4 g	Chipotle Peppers with adobo	Recipe .4 g	Total Grams	Total Amino Acid	Total Protein/g
Essential Amino Acids		Protein/g	Recipe	Protein/g	Recipe			
Histidine	18	23.03	124.38	20.00	8.00	5.80	132.38	22.82
Isoleuciine	25	50.46	272.49	33.00	13.20		285.69	49.26
Methionine	25	21.00	113.41	13.00	5.20		118.61	20.45
Leucine	55	84.79	457.86	53.00	21.20		479.06	82.60
Lysine	51	69.55	375.57	46.00	18.40		393.97	67.93
Phenylalanine	47	42.61	230.12	32.00	12.80		242.92	41.88
Threonine	27	37.93	204.83	36.50	14.60		219.43	37.83
Tryptophan	7	12.20	65.86	13.00	5.20		71.06	12.25
Valine	32	57.85	312.41	43.00	17.20		329.61	56.83
Recipe Amount		1 Cup		2 peppers				
Protein per recipe (g)		5.40		0.40				

Total recipe Protein (g)	# of servings	Protein / serving (g)	Complete Protein
5.80	16.00	0.36	

Cilantro Lime Crema

Now let's add some of that extra flavor we spoke about above. Here we make the basic crema as above. After the 8 - 24 hour waiting period, we add the flavoring before sealing and placing in the fridge. The flavors will absorb into the fat the longer it sits in the fridge, but you can use it right away.

Ingredients:

- 1 cup heavy whipping cream
- 1 tablespoon buttermilk
- 1 tablespoon cilantro
- 1 tablespoon lime juice and zest of one lime

Total time: 24 hours Yield: 1 cup Portions: 16

Serving size: 1 Tbsp

Food Science Note

Sour cream, creme fraiche, and buttermilk are examples of cultured milk products.These are milk products with bacteria that produce lactic acid. This acid lowers the ph and fends off the the bad microbes.

The idea of leaving a milk based product at room temperature is counterintuitive. According to McGee, (2103, p. 31), milk is a rich source of nutrients, for both animals and microbes. Raw milk allows for acid producing bacteria to start earlier and thrive over the less desirable microbes.

When we pasteurize milk, we kill most of the good acid producing bacteria, leaving the less desirable microbes to take over and spoil the milk product.

By adding the cultured buttermilk to the pasteurized milk we are adding the acid producers back into the program. The acid in the milk gives it a pleasing sourness, but if allowed to continue too far, the taste and safety of the product changes.

The time at room temperature allows for the pleasing sourness to thrive without going beyond an acceptable limit. Refrigeration will slow down the production of any harmful microbes once the acid is produced.

Cilantro Lime Crema

		Fat (g)	Sat. Fat (g)	Phos (mg)	Pot (mg)	Sodium (mg)	Calories	Carb (g)	Vit A (iu)	Vit C (mg)	Vit K (Mcg)	Vit E (mg) AT	Vit B6 (mg)
	RDA	**65**	**20**	**700**	**3300**	**2300**	**2000**	**275**	**5000**	**60**	**120**	**20**	**2**
Ingredients	**Amt**												
Creme Fraiche	**1 cup**	88.14	55.08	161.63	201.13	106.46	827.13	7.75	3502.98	1.40	7.61	2.51	0.11
		135.60%	275.41%	23.09%	6.09%	4.63%	41.36%	2.82%	70.06%	2.33%	6.34%	12.53%	5.31%
Green Onion	**1 Tbsp**	0.01	0.00	2.31	17.25	1.00	2.00	0.44	62.31	1.18	12.94	12.94	0.00
		0.02%	0.00%	0.33%	0.52%	0.04%	0.10%	0.16%	1.25%	64.80%	10.78%	64.69%	0.00%
Cilantro, fresh	**1 Tbsp**	0.30	0.00	20.10	219.00	2.15	9.60	1.50	944.58	11.40	130.20	1.05	0.00
				2.87%									
Lime juice and zest (1 lime)	**1 Tbsp**	0.00	0.00	3.99	32.80	0.50	7.00	2.00	14.00	8.40	0.20	0.10	0.00
Total for Dish		88.45	55.08	188.03	470.18	110.11	845.73	11.69	4523.88	22.38	150.95	16.59	0.11
		136.08%	275.41%	26.86%	14.25%	4.79%	42.29%	4.25%	90.48%	37.29%	125.79%	82.97%	5.31%
# of Servings		16	16	16	16	16	16	16	16	16	16	16	16
(1 Tbsp)													
Totals per serving		5.53	3.44	11.75	29.39	6.88	52.86	0.73	282.74	1.40	9.43	1.04	0.01
		8.50%	17.21%	1.68%	0.89%	0.30%	2.64%	0.27%	5.65%	2.33%	7.86%	5.19%	0.33%

Essential Amino Acid Chart for recipe Lime Cilantro

	Mg/ gram protein needed	Creme Fraiche	Recipe 5.4 g	Green onion	Recipe .11 g	Lime juice	Recipe .5 g	Total Grams	Total Amino Acid	Total Protein/g
Essential Amino Acids		Protein/g	Recipe	Protein/g	Recipe	Protein/g	Recipe			
Histidine	18	23.03	124.38	17.78	2.00	6.00	3.00	6.01	129.38	21.52
Isoleuciine	25	50.46	272.49	42.78	4.81	6.00	3.00		280.30	46.62
Methionine	25	21.00	113.41	11.11	1.25	6.00	3.00		117.66	19.57
Leucine	55	84.79	457.86	60.56	6.81	45.00	22.50		487.17	81.03
Lysine	51	69.55	375.57	50.56	5.69	45.00	22.50		403.76	67.15
Phenylalanine	47	42.61	230.12	32.78	3.69	31.00	15.50		249.31	41.46
Threonine	27	37.93	204.83	40.00	4.50	6.00	3.00		212.33	35.32
Tryptophan	7	12.20	65.86	11.11	1.25	6.00	3.00		70.11	11.66
Valine	32	57.85	312.41	45.00	5.06	31.00	15.50		332.97	55.38
Recipe Amount		1 cup		1 Tbsp		1 Tbsp				
Protein per recipe (g)		5.40		0.11		0.50				

Total recipe Protein (g)	# of servings	Protein / serving (g)	Complete Protein
6.01	16.00	0.38	

Chicken Thighs

Chicken thighs and drumsticks have more flavor than chicken breasts. Whether or not you like that flavor is another matter. The higher fat content contributes to that flavor.

Because the fat content is higher it should be taken into consideration when deciding on a protein choice. You certainly could use chicken breasts, but be cautious not to overcook.

Ingredients:

- 2 chicken thighs
- 2 garlic cloves
- 2 tablespoons thyme
- 2 tablespoons olive oil
- 1 dried ancho chili pepper (or other mild pepper)
- 1 dried arbol pepper (or other hot pepper)
- 1 tablespoon cumin
- 1/2 teaspoon kosher salt
- 1/2 teaspoon black pepper
- 1 cup water

Heat up a cast iron skillet or other heavy bottomed pan on medium high heat. When the pan in hot, add the oil. Brown the chicken thighs, skin side down first, and then flip, about two minutes per side.

Add the rest of the ingredients to the pan, adjusting the water so no more than half the meat is covered. Cover the pan and lower the heat to medium until the temp of the thighs reaches 165° F, about 45 minutes.

You may need to add some water if it gets too low, but don't submerge the thighs.

When done, remove the thighs to a cutting surface and use two forks to shred the meat. If you used bone in and skin on thighs, you will want to remove those from the final product. They can also be eaten as a meal by themselves.

I like to turn up the heat on what is remaining in the skillet and reduce the liquid and other ingredients into a thicker sauce to mix into the chicken for added flavor and moisture.

The other cooking option is to place all the ingredients in a pot large enough to hold all the chicken and submerge with water. Cook for the same amount of time and prepare the same way.

To add a peppery or BBQ flavor, simple mix 2 Tbsp of our chili paste or BBQ sauce after the meat is shredded and mix thoroughly.

This last step is optional and may not give you the lowest level of phosphorus possible, but it will increase flavor.

Food Note

In "*The Effect of Various Boiling Conditions on Reduction of Phosphorus and Protein in Meat*" (S. Ando et. al., 2015), the authors conclude that the style of cooking meat can have an effect on certain nutrients in the final product. If you boil your meat, for instance, some of the phosphorus will leach out into the cooking water, and thus reduce the total phosphorus when eaten.

Since the idea of eating boiled meat is pretty unappealing, braising is a much better way to approach the science. Boiling is completely submerging the food into water. Stewing is similar but is done with smaller cuts of meat, and in most cases, consuming the dish with the liquid. Braising is cooking in less amounts of water and at lower temperatures. The liquid is often served with the meat, similar to stewing.

If you braise your meat, and then discard the liquid you will minimize the phosphorus in the food. Unfortunately lots of the flavor, along with the phosphorus, is in that liquid and it is very tempting to use. Rarely are the remnants of boiling considered appetizing or ever used in a final product, with the exception of stock or broth.

Fish Tacos

Strong toppings can hide the fish flavor and this may or may not be what you are looking for. Keep the flavor of the fish in mind when choosing how to build your tacos.

To boost the flavor of a mild fish, I like to marinate it for an hour or so, and then pan fry in strips to fit best in the taco. You can use this marinating method for chicken, fish, and pork.

In addition, if you look through the phosphorus/protein chart in the back of this book, you will notice that some of lowest ratios appear in the seafood section, making fish tacos a good choice.

Ingredients:

- 2 pounds haddock, tilapia, or cod, cut into strips
- 1/4 cup plus 1 tablespoon olive oil
- 1 tablespoon shallot, diced
- 1 lime, juice and zest
- 1/2 teaspoon cumin
- 1 garlic clove, smashed
- 1 tablespoon white wine vinegar
- 1/2 teaspoon oregano
- 1/2 teaspoon kosher salt

Total time: 1 hour, 15 min. Yield: 1.5 lbs. Portions: 8

Serving size: 3 oz.

Mix all the ingredients minus 1 tablespoon of olive oil and place the fish and marinade in a plastic bag. Refrigerate for 1 hour.

Heat a pan on medium heat. When hot, add 1 tablespoon olive oil and cook the fish for 2 minutes, or until opaque in color. When done, remove the fish and place it on a cutting board and carefully shred the meat with a fork.

Tacos in Different Combinations

Now that all the ingredients have been produced, we can start to put together different combinations to make the final product. We started out by making our own tortillas because of the high sodium content in store bought options. The finished tortillas above range from 3.8% to 6.8% sodium total per taco. Recall that each store bought tortilla has 15% sodium each, and that is just for the shell.

The different recipes below also point out the variations we have in putting together a taco with the associated nutrition and protein charts. Some are higher in fat and saturated fat, some are higher in phosphorus, some in potassium. The point is that when a dietitian answers the very common question "Can I eat that?" with the response ,"Well, it depends," they are correct. It depends on what you put into your taco, how it is made, and what are the danger areas for each individual.

One of the most important parts of participating in your own care is getting to know what is going on with your body. One way of doing this is to pay attention to your blood work, and understand what those numbers mean. Working together with the dietitian and applying the detailed information in each recipe can help you choose the best options for your particular situation.

It is also important to point out that the flour tortilla recipe used contains lard, which adds 4.8 grams (7.38%) of fat, and 1.88 grams (9.38%) of saturated fat for the total recipe. Simply make the flour tortillas without the lard to reduce both numbers significantly in the final product.

Corn Taco, Chicken, Pico de Gallo, Crema

		Fat (g)	Sat. Fat (g)	Phos (mg)	Pot (mg)	Sodium (mg)	Calories	Carb (g)	Vit A (iu)	Vit C (mg)	Vit K (Mcg)	Vit E (mg) AT	Vit B6 (mg)
	RDA	65	20	700	3300	2300	2000	275	5000	60	120	20	2
Ingredients	Amt												
1 Corn Tortilla	1.00	0.67	0.17	42.33	56.67	128.74	69.33	14.56	40.67	0.00	0.00	0.00	0.07
		1.03%	0.83%	6.05%	1.72%	5.60%	3.47%	5.29%	0.81%	0.00%	0.00%	0.00%	3.33%
Chicken Thighs	1.5 oz	4.05	1.20	63.50	76.80	31.00	82.10	0.00	26.10	0.00	1.50	0.15	0.15
		6.23%	6.00%	9.07%	2.33%	1.35%	4.11%	0.00%	0.52%	0.00%	1.25%	0.75%	7.50%
Pico de Gallo	2 Tbsp	0.04	0.00	10.24	86.13	50.88	8.83	2.12	409.69	5.66	13.43	0.19	0.02
		0.06%	0.00%	1.46%	2.61%	2.21%	0.44%	0.77%	8.19%	9.43%	11.19%	0.97%	0.94%
Lime Cliantro Crema	1 Tbsp	5.55	3.44	13.46	46.20	7.11	53.97	0.98	346.55	2.71	18.39	1.92	0.07
		8.54%	17.21%	1.92%	1.40%	0.31%	2.70%	0.35%	6.93%	4.52%	15.33%	9.59%	3.44%
Total for Dish		10.31	4.81	129.53	265.79	217.73	214.23	17.65	823.00	8.37	33.32	2.26	0.30
		15.86%	24.05%	18.50%	8.05%	9.47%	10.71%	6.42%	16.46%	13.95%	27.77%	11.30%	15.21%
1 Taco	1	1	1	1	1	1	1	1	1	1	1	1	1
Totals per serving		10.31	4.81	129.53	265.79	217.73	214.23	17.65	823.00	8.37	33.32	2.26	0.30
		15.86%	24.05%	18.50%	8.05%	9.47%	10.71%	6.42%	16.46%	13.95%	27.77%	11.30%	15.21%

Essential Amino Acid Chart for recipe Corn, Chicken, Pico de Gallo, Crema

	Mg/ gram protein needed	Chicken Thigh	Recipe 10.5 g	Pico de Gallo	Recipe .32 g	Lime Cilantro Crema	Recipe .35 g	Corn Tortilla	Recipe 1.2 g	Total Gram	Total Amino Acid	Total Protein/g
Essential Amino Acids		Protein/g	Recipe	Protein/g	Recipe	Protein/g	Recipe	Protein/g	Recipe			
Histidine	18	31.00	325.50	16.81	5.38	38.70	13.55	30.66	36.79	12.37	381.22	30.82
Isoleuciine	25	52.86	555.00	29.41	9.41	53.00	18.55	35.94	43.13		626.09	50.61
Methionine	25	27.71	291.00	43.81	14.02	26.80	9.38	21.04	25.25		339.64	27.46
Leucine	55	75.00	787.50	41.72	13.35	96.70	33.85	123.21	147.85		982.54	79.43
Lysine	51	84.86	891.00	39.16	12.53	92.58	32.40	28.30	33.96		969.90	78.41
Phenylalanine	47	39.71	417.00	57.51	18.40	53.80	18.83	49.34	59.21		513.44	41.51
Threonine	27	42.29	444.00	33.60	10.75	36.90	12.92	37.74	45.28		512.95	41.47
Tryptophan	7	11.69	122.70	8.51	2.72	13.50	4.73	7.09	8.51		138.66	11.21
Valine	32	49.57	520.50	30.51	9.76	68.70	24.05	50.85	61.02		615.33	49.74
Recipe Amount		1.5 oz		2 Tbsp				1				
Protein per recipe (g)	12.37	10.5		0.32		0.35		1.2				

Total recipe Protein (g)	# of servings	Protein / serving (g)	Complete Protein
12.37	1.00	12.37	

Corn Taco, Chicken Thighs, Pico de gallo

		Fat (g)	Sat. Fat (g)	Phos (mg)	Pot (mg)	Sodium (mg)	Calories	Carb (g)	Vit A (iu)	Vit C (mg)	Vit K (Mcg)	Vit E (mg) AT	Vit B6 (mg)
	RDA	**65**	**20**	**700**	**3300**	**2300**	**2000**	**275**	**5000**	**60**	**120**	**20**	**2**
Ingredients	**Amt**												
1 Corn Tortilla	**1.00**	0.67	0.17	42.33	56.67	128.74	69.33	14.56	40.67	0.00	0.00	0.00	0.07
		1.03%	0.83%	6.05%	1.72%	5.60%	3.47%	5.29%	0.81%	0.00%	0.00%	0.00%	3.33%
Chicken Thighs	**1.5 oz**	4.05	1.20	63.50	76.80	31.00	82.10	0.00	26.10	0.00	1.50	0.15	0.15
		6.23%	6.00%	9.07%	2.33%	1.35%	4.11%	0.00%	0.52%	0.00%	1.25%	0.75%	7.50%
Pico de Gallo	**2 Tbsp**	0.04	0.00	10.24	86.13	50.88	8.83	2.12	409.69	5.66	13.43	0.19	0.02
		0.06%	0.00%	1.46%	2.61%	2.21%	0.44%	0.77%	8.19%	9.43%	11.19%	0.97%	0.94%
Total for Dish		4.76	1.37	116.07	219.59	210.62	160.27	16.67	476.45	5.66	14.93	0.34	0.24
		7.32%	6.83%	16.58%	6.65%	9.16%	8.01%	6.06%	9.53%	9.43%	12.44%	1.72%	11.77%
1Taco	**1**	1	1	1	1	1	1	1	1	1	1	1	1
Totals per serving		4.76	1.37	116.07	219.59	210.62	160.27	16.67	476.45	5.66	14.93	0.34	0.24
	8 6/7	7.32%	6.83%	16.58%	6.65%	9.16%	8.01%	6.06%	9.53%	9.43%	12.44%	1.72%	11.77%

Essential Amino Acid Chart for Corn Taco, Chicken, Pico de Gallo

	Mg/ gram protein needed	Chicken Thigh	Recipe 10.5 g	Pico de Gallo	Recipe .32 g	Corn Tortila	Recipe 1.2 g	Total Grams	Total Amino Acid	Total Protein/g
Essential Amino Acids		Protein/g	Recipe	Protein/g	Recipe	Protein/g	Recipe			
Histidine	18	31.00	325.50	16.81	5.38	30.66	36.79	12.02	367.67	30.59
Isoleuciine	25	52.86	555.00	29.41	9.41	35.94	43.13		607.54	50.54
Methionine	25	27.71	291.00	43.81	14.02	21.04	25.25		330.26	27.48
Leucine	55	75.00	787.50	41.72	13.35	123.21	147.85		948.70	78.93
Lysine	51	84.86	891.00	39.16	12.53	28.30	33.96		937.49	77.99
Phenylalanine	47	39.71	417.00	57.51	18.40	49.34	59.21		494.61	41.15
Threonine	27	42.29	444.00	33.60	10.75	37.74	45.28		500.04	41.60
Tryptophan	7	11.69	122.70	8.51	2.72	7.09	8.51		133.94	11.14
Valine	32	49.57	520.50	30.51	9.76	50.85	61.02		591.28	49.19
Recipe Amount		1.5 oz		2 Tbsp		1				
Protein per recipe (g)	12.02	10.5		0.32		1.2				

Total recipe Protein (g)	# of servings	Protein / serving (g)	Complete Protein
12.02	1.00	12.02	

Flour Taco, Chicken Thighs, Pico de Gallo, Crema

		Fat (g)	Sat. Fat (g)	Phos (mg)	Pot (mg)	Sodium (mg)	Calories	Carb (g)	Vit A (iu)	Vit C (mg)	Vit K (Mcg)	Vit E (mg) AT	Vit B6 (mg)
	RDA	65	20	700	3300	2300	2000	275	5000	60	120	20	2
Ingredients	Amt												
1 Flour Tortilla	1.00	5.10	1.93	33.75	33.50	144.38	169.38	16.38	0.63	0.00	0.10	0.08	0.03
		7.85%	9.63%	4.82%	1.02%	6.28%	8.47%	5.95%	0.01%	0.00%	0.08%	0.38%	1.25%
Chicken Thighs	1.5 oz	4.05	1.20	63.50	76.80	31.00	82.10	0.00	26.10	0.00	1.50	0.15	0.15
		6.23%	6.00%	9.07%	2.33%	1.35%	4.11%	0.00%	0.52%	0.00%	1.25%	0.75%	7.50%
Pico de Gallo	2 Tbsp	0.04	0.00	10.24	86.13	50.88	8.83	2.12	409.69	5.66	13.43	0.19	0.02
		0.06%	0.00%	1.46%	2.61%	2.21%	0.44%	0.77%	8.19%	9.43%	11.19%	0.97%	0.94%
Chipotle Crema	1 Tbsp	5.51	3.44	11.16	23.20	98.65	53.14	0.73	309.94	0.64	1.18	0.21	0.07
		8.48%	17.21%	1.59%	0.70%	4.29%	2.66%	0.27%	6.20%	1.06%	0.98%	1.03%	3.44%
Total for Dish		14.70	6.57	118.65	219.62	324.90	313.45	19.23	746.35	6.30	16.20	0.63	0.26
		22.62%	32.84%	16.95%	6.66%	14.13%	15.67%	6.99%	14.93%	10.49%	13.50%	3.13%	13.13%
1 Taco		1	1	1	1	1	1	1	1	1	1	1	1
Totals per serving		14.70	6.57	118.65	219.62	324.90	313.45	19.23	746.35	6.30	16.20	0.63	0.26
		22.62%	32.84%	16.95%	6.66%	14.13%	15.67%	6.99%	14.93%	10.49%	13.50%	3.13%	13.13%

Essential Amino Acid Chart for recipe Flour, Chicken, Pico De Gallo, Crema

	Mg/ gram protein needed	Chicken Thigh	Recipe 10.5 g	Pico de Gallo	Recipe .32 g	Chipotle Crema	Recipe .35 g	Corn Tortila	Recipe .8 g	Total Gram	Total Amino Acid	Total Protein/g
Essential Amino Acids		Protein/g	Recipe	Protein/g	Recipe	Protein/g	Recipe	Protein/g	Recipe			
Histidine	18	31.00	325.50	16.81	5.38	0.24	0.08	22.33	17.86	11.97	348.82	29.14
Isoleuciine	25	52.86	555.00	29.41	9.41	0.42	0.15	34.57	27.66		592.22	49.48
Methionine	25	27.71	291.00	43.81	14.02	0.63	0.22	17.75	14.20		319.44	26.69
Leucine	55	75.00	787.50	41.72	13.35	0.60	0.21	68.84	55.07		856.13	71.52
Lysine	51	84.86	891.00	39.16	12.53	0.56	0.20	22.09	17.67		921.40	76.98
Phenylalanine	47	39.71	417.00	57.51	18.40	0.82	0.29	50.39	40.31		476.00	39.77
Threonine	27	42.29	444.00	33.60	10.75	0.48	0.17	27.21	21.77		476.69	39.82
Tryptophan	7	11.69	122.70	8.51	2.72	0.12	0.04	12.33	9.86		135.33	11.31
Valine	32	49.57	520.50	30.51	9.76	0.44	0.15	40.23	32.19		562.60	47.00
Recipe Amount		1.5 oz		2 Tbsp				1				
Protein per recipe (g)	11.97	10.5		0.32		0.35		0.8				

Total recipe Protein (g)	# of servings	Protein / serving (g)	Complete Protein
11.97	1.00	11.97	

Corn Taco, Cod, Pico de Gallo, Crema

		Fat (g)	Sat. Fat (g)	Phos (mg)	Pot (mg)	Sodium (mg)	Calorie	Carb (g)	Vit A (iu)	Vit C (mg)	Vit K (Mcg)	Vit E (mg) AT	Vit B6 (mg)
	RDA	65	20	700	3300	2300	2000	275	5000	60	120	20	2
Ingredients	Amt												
1 Corn Tortilla	1.00	0.67	0.17	42.33	56.67	128.74	69.33	14.56	40.67	0.00	0.00	0.00	0.07
		1.03%	0.83%	6.05%	1.72%	5.60%	3.47%	5.29%	0.81%	0.00%	0.00%	0.00%	3.33%
Cod	1.5 oz	0.30	0.00	57.90	102.45	32.70	44.10	0.00	20.25	0.45	0.00	0.30	0.30
		0.46%	0.00%	8.27%	3.10%	1.42%	2.21%	0.00%	0.41%	0.75%	0.00%	1.50%	15.00%
Pico de Gallo	2 Tbsp	0.04	0.00	10.24	86.13	50.88	8.83	2.12	409.69	5.66	13.43	0.19	0.02
		0.06%	0.00%	1.46%	2.61%	2.21%	0.44%	0.77%	8.19%	9.43%	11.19%	0.97%	0.94%
Lime Cliantro Crema	1 Tbsp	5.55	3.44	13.46	46.20	7.11	53.97	0.98	346.55	2.71	18.39	1.92	0.07
		8.54%	17.21%	1.92%	1.40%	0.31%	2.70%	0.35%	6.93%	4.52%	15.33%	9.59%	3.44%
Total for Dish		6.56	3.61	123.93	291.44	219.43	176.23	17.65	817.15	8.82	31.82	2.41	0.45
		10.09%	18.05%	17.70%	8.83%	9.54%	8.81%	6.42%	16.34%	14.70%	26.52%	12.05%	22.71%
1 Taco	1	1	1	1	1	1	1	1	1	1	1	1	1
Totals per serving		6.56	3.61	123.93	291.44	219.43	176.23	17.65	817.15	8.82	31.82	2.41	0.45
	8 6/7	10.09%	18.05%	17.70%	8.83%	9.54%	8.81%	6.42%	16.34%	14.70%	26.52%	12.05%	22.71%

Essential Amino Acid Chart for recipe Corn Tortilla, Cod, Pico de Gallo, Crema

	Mg/ gram protein needed	Cod	Recipe 9.6 g	Pico de Gallo	Recipe .32 g	Lime Cilantro Crema	Recipe .35 g	Corn Tortila	Recipe 1.2 g	Total Grams	Total Amino Acid	Total Protein/g
Essential Amino Acids		Protein/g	Recipe	Protein/g	Recipe	Protein/g	Recipe	Protein/g	Recipe			
Histidine	18	29.38	282.00	16.81	5.38	38.70	13.55	30.66	36.79	11.47	337.72	29.44
Isoleuciine	25	44.53	427.50	29.41	9.41	53.00	18.55	35.94	43.13		498.59	43.47
Methionine	25	29.53	283.50	43.81	14.02	26.80	9.38	21.04	25.25		332.14	28.96
Leucine	55	81.25	780.00	41.72	13.35	96.70	33.85	123.21	147.85		975.04	85.01
Lysine	51	91.72	880.50	39.16	12.53	92.58	32.40	28.30	33.96		959.40	83.64
Phenylalanine	47	38.91	373.50	57.51	18.40	53.80	18.83	49.34	59.21		469.94	40.97
Threonine	27	43.75	420.00	33.60	10.75	36.90	12.92	37.74	45.28		488.95	42.63
Tryptophan	7	11.20	107.55	8.51	2.72	13.50	4.73	7.09	8.51		123.51	10.77
Valine	32	51.41	493.50	30.51	9.76	68.70	24.05	50.85	61.02		588.33	51.29
Recipe Amout		1.5 oz		2 Tbsp		1 Tbsp		1				
Protein per recipe (g)	11.47	9.6		0.32		0.35		1.2				

Total recipe Protein (g)	# of servings	Protein / serving (g)	Complete Protein
11.47	1.00	11.47	

Corn Taco, Cod, Mango Salsa, Crema

		Fat (g)	Sat. Fat (g)	Phos (mg)	Pot (mg)	Sodium (mg)	Calories	Carb (g)	Vit A (iu)	Vit C (mg)	Vit K (Mcg)	Vit E (mg) AT	Vit B6 (mg)
	RDA	65	20	700	3300	2300	2000	275	5000	60	120	20	2
Ingredients	Amt												
1 Corn Tortilla	1.00	0.67	0.17	42.33	56.67	128.74	69.33	14.56	40.67	0.00	0.00	0.00	0.07
		1.03%	0.83%	6.05%	1.72%	5.60%	3.47%	5.29%	0.81%	0.00%	0.00%	0.00%	3.33%
Cod	1.5 oz	0.30	0.00	57.90	102.45	32.70	44.10	0.00	20.25	0.45	0.00	0.30	0.30
		0.46%	0.00%	8.27%	3.10%	1.42%	2.21%	0.00%	0.41%	0.75%	0.00%	1.50%	15.00%
Mango Salso	2 Tbsp	0.05	0.01	4.76	47.50	25.25	12.19	6.79	398.59	13.10	6.78	0.30	0.04
		0.08%	0.04%	0.68%	1.44%	1.10%	0.61%	2.47%	7.97%	21.84%	5.65%	1.48%	2.14%
Lime Cliantro Crema	1 Tbsp	5.55	3.44	13.46	46.20	7.11	53.97	0.98	346.55	2.71	18.39	1.92	0.07
		8.54%	17.21%	1.92%	1.40%	0.31%	2.70%	0.35%	6.93%	4.52%	15.33%	9.59%	3.44%
Total for Dish		6.57	3.62	118.46	252.81	193.80	179.59	22.32	806.06	16.26	25.17	2.51	0.48
		10.11%	18.09%	16.92%	7.66%	8.43%	8.98%	8.12%	16.12%	27.11%	20.97%	12.57%	23.91%
1 Taco	1	1	1	1	1	1	1	1	1	1	1	1	1
Totals per serving		6.57	3.62	118.46	252.81	193.80	179.59	22.32	806.06	16.26	25.17	2.51	0.48
		10.11%	18.09%	16.92%	7.66%	8.43%	8.98%	8.12%	16.12%	27.11%	20.97%	12.57%	23.91%

Essential Amino Acid Chart for recipe Corn, Cod, Mango Salsa, Crema

	Mg/ gram protein needed	Cod, Dry heat	Recipe 9.6 g	Pico de Gallo	Recipe .32 g	Lime Cilantro Crema	Recipe .35 g	Corn Tortila	Recipe 1.2 g	Total Gram	Total Amino Acid	Total Protein/g
Essential Amino Acids		Protein/g	Recipe	Protein/g	Recipe	Protein/g	Recipe	Protein/g	Recipe			
Histidine	18	29.38	282.00	16.81	5.38	38.70	13.55	30.66	36.79	11.47	337.72	29.44
Isoleuciine	25	44.53	427.50	29.41	9.41	53.00	18.55	35.94	43.13		498.59	43.47
Methionine	25	29.53	283.50	43.81	14.02	26.80	9.38	21.04	25.25		332.14	28.96
Leucine	55	81.25	780.00	41.72	13.35	96.70	33.85	123.21	147.85		975.04	85.01
Lysine	51	91.72	880.50	39.16	12.53	92.58	32.40	28.30	33.96		959.40	83.64
Phenylalanine	47	38.91	373.50	57.51	18.40	53.80	18.83	49.34	59.21		469.94	40.97
Threonine	27	43.75	420.00	33.60	10.75	36.90	12.92	37.74	45.28		488.95	42.63
Tryptophan	7	11.20	107.55	8.51	2.72	13.50	4.73	7.09	8.51		123.51	10.77
Valine	32	51.41	493.50	30.51	9.76	68.70	24.05	50.85	61.02		588.33	51.29
Recipe Amount		1.5 oz		2 Tbsp				1				
Protein per recipe (g)	11.47	9.6		0.32		0.35		1.2				

Total recipe Protein (g)	# of servings	Protein / serving (g)	Complete Protein
11.47	1.00	11.47	

Flour Taco, Cod, Mango Salsa, Crema-1

		Fat (g)	Sat. Fat (g)	Phos (mg)	Pot (mg)	Sodiu m (mg)	Calorie s	Carb (g)	Vit A (iu)	Vit C (mg)	Vit K (Mcg)	Vit E (mg) AT	Vit B6 (mg)
	RDA	**65**	**20**	**700**	**3300**	**2300**	**2000**	**275**	**5000**	**60**	**120**	**20**	**2**
Ingredients	**Amt**												
1 Flour Tortilla	**1.00**	5.10	1.93	33.75	33.50	144.38	169.38	16.38	0.63	0.00	0.10	0.08	0.03
		7.85%	9.63%	4.82%	1.02%	6.28%	8.47%	5.95%	0.01%	0.00%	0.08%	0.38%	1.25%
Cod	**1.5 oz**	0.30	0.00	57.90	102.45	32.70	44.10	0.00	20.25	0.45	0.00	0.30	0.30
		0.46%	0.00%	8.27%	3.10%	1.42%	2.21%	0.00%	0.41%	0.75%	0.00%	1.50%	15.00%
Mango Salso	**2 Tbsp**	0.05	0.01	4.76	47.50	25.25	12.19	6.79	398.59	13.10	6.78	0.30	0.04
		0.08%	0.04%	0.68%	1.44%	1.10%	0.61%	2.47%	7.97%	21.84%	5.65%	1.48%	2.14%
Lime Cliantro Crema	**1 Tbsp**	5.55	3.44	13.46	46.20	7.11	53.97	0.98	346.55	2.71	18.39	1.92	0.07
		8.54%	17.21%	1.92%	1.40%	0.31%	2.70%	0.35%	6.93%	4.52%	15.33%	9.59%	3.44%
Total for Dish		11.00	5.38	109.87	229.65	209.43	279.63	24.14	766.02	16.26	25.27	2.59	0.44
		16.93%	26.88%	15.70%	6.96%	9.11%	13.98%	8.78%	15.32%	27.11%	21.06%	12.95%	21.82%
1 Taco	**1**	1	1	1	1	1	1	1	1	1	1	1	1
Totals per serving		11.00	5.38	109.87	229.65	209.43	279.63	24.14	766.02	16.26	25.27	2.59	0.44
	8 6/7	16.93%	26.88%	15.70%	6.96%	9.11%	13.98%	8.78%	15.32%	27.11%	21.06%	12.95%	21.82%

Essential Amino Acid Chart for recipe Flour, Cod, Mango Salsa, Crema

	Mg/ gram protein needed	Cod, Dry heat	Recipe 9.6 g	Pico de Gallo	Recipe 3.2 g	Lime Cilantro Crema	Recipe . 35 g	Flour Tortila	Recipe .8 g	Total Gram	Total Amino Acid	Total Protein/g
Essential Amino Acids		Protein/g	Recipe	Protein/g	Recipe	Protein/ g	Recipe	Protein/g	Recipe			
Histidine	18	29.38	282.00	16.81	5.38	38.70	13.55	22.33	17.86	11.07	318.78	28.80
Isoleuciine	25	44.53	427.50	29.41	9.41	53.00	18.55	34.57	27.66		483.12	43.64
Methionine	25	29.53	283.50	43.81	14.02	26.80	9.38	17.75	14.20		321.10	29.01
Leucine	55	81.25	780.00	41.72	13.35	96.70	33.85	68.84	55.07		882.26	79.70
Lysine	51	91.72	880.50	39.16	12.53	92.58	32.40	22.09	17.67		943.11	85.20
Phenylalanine	47	38.91	373.50	57.51	18.40	53.80	18.83	50.39	40.31		451.04	40.74
Threonine	27	43.75	420.00	33.60	10.75	36.90	12.92	27.21	21.77		465.44	42.04
Tryptophan	7	11.20	107.55	8.51	2.72	13.50	4.73	12.33	9.86		124.86	11.28
Valine	32	51.41	493.50	30.51	9.76	68.70	24.05	40.23	32.19		559.49	50.54
Recipe Amount		1.5 oz		2 Tbsp		1 Tbsp		1				
Protein per recipe (g)	11.07	9.6		0.32		0.35		0.8				

Total recipe Protein (g)	# of servings	Protein / serving (g)	Complete Protein
11.07	1.00	11.07	

Flour Taco, Cod, Mango Salsa, Crema

		Fat (g)	Sat. Fat (g)	Phos (mg)	Pot (mg)	Sodium (mg)	Calories	Carb (g)	Vit A (iu)	Vit C (mg)	Vit K (Mcg)	Vit E (mg) AT	Vit B6 (mg)
	RDA	65	20	700	3300	2300	2000	275	5000	60	120	20	2
Ingredients	Amt												
1 Flour Tortilla	1.00	5.10	1.93	33.75	33.50	144.38	169.38	16.38	0.63	0.00	0.10	0.08	0.03
		7.85%	9.63%	4.82%	1.02%	6.28%	8.47%	5.95%	0.01%	0.00%	0.08%	0.38%	1.25%
Cod	1.5 oz	0.30	0.00	57.90	102.45	32.70	44.10	0.00	20.25	0.45	0.00	0.30	0.30
		0.46%	0.00%	8.27%	3.10%	1.42%	2.21%	0.00%	0.41%	0.75%	0.00%	1.50%	15.00%
Mango Salso	2 Tbsp	0.05	0.01	4.76	47.50	25.25	12.19	6.79	398.59	13.10	6.78	0.30	0.04
		0.08%	0.04%	0.68%	1.44%	1.10%	0.61%	2.47%	7.97%	21.84%	5.65%	1.48%	2.14%
Lime Cliantro Crema	1 Tbsp	5.55	3.44	13.46	46.20	7.11	53.97	0.98	346.55	2.71	18.39	1.92	0.07
		8.54%	17.21%	1.92%	1.40%	0.31%	2.70%	0.35%	6.93%	4.52%	15.33%	9.59%	3.44%
Total for Dish		11.00	5.38	109.87	229.65	209.43	279.63	24.14	766.02	16.26	25.27	2.59	0.44
		16.93%	26.88%	15.70%	6.96%	9.11%	13.98%	8.78%	15.32%	27.11%	21.06%	12.95%	21.82%
1 Taco	1	1	1	1	1	1	1	1	1	1	1	1	1
Totals per serving		11.00	5.38	109.87	229.65	209.43	279.63	24.14	766.02	16.26	25.27	2.59	0.44
	8 6/7	16.93%	26.88%	15.70%	6.96%	9.11%	13.98%	8.78%	15.32%	27.11%	21.06%	12.95%	21.82%

Essential Amino Acid Chart for recipe Flour, Cod, Mango Salsa

	Mg/ gram protein needed	Cod, Dry heat	Recipe 9.6 g	Pico de Gallo	Recipe .32 g	Flour Tortila	Recipe .8 g	Total Grams	Total Amino Acid	Total Protein/g
Essential Amino Acids		Protein/g	Recipe	Protein/g	Recipe	Protein/g	Recipe			
Histidine	18	29.38	282.00	16.81	5.38	22.33	17.86	10.72	305.24	28.47
Isoleuciine	25	44.53	427.50	29.41	9.41	34.57	27.66		464.57	43.34
Methionine	25	29.53	283.50	43.81	14.02	17.75	14.20		311.72	29.08
Leucine	55	81.25	780.00	41.72	13.35	68.84	55.07		848.42	79.14
Lysine	51	91.72	880.50	39.16	12.53	22.09	17.67		910.71	84.95
Phenylalanine	47	38.91	373.50	57.51	18.40	50.39	40.31		432.21	40.32
Threonine	27	43.75	420.00	33.60	10.75	27.21	21.77		452.52	42.21
Tryptophan	7	11.20	107.55	8.51	2.72	12.33	9.86		120.13	11.21
Valine	32	51.41	493.50	30.51	9.76	40.23	32.19		535.45	49.95
Recipe Amount		1.5 oz		2 Tbsp		1				
Protein per recipe (g)		9.6		0.32		0.8				

Total recipe Protein (g)	# of servings	Protein / serving (g)	Complete Protein
10.72	1.00	10.72	

Spices for more Flavor

Using spices to elevate flavor and reduce the amount of sodium is a part of chronic kidney disease dietary programs. Salt in food has two components, it has a flavor all it's own and it enhances the overall taste of the final dish. As mentioned in other sections, salt will open up taste buds on the tongue that do not generally open up for other flavors. Once open with salt, they can then receive other flavors, giving food more access to taste buds that do not normally accept those flavors.

When food tastes bland, the tongue is not receiving all the flavors available or there is not enough flavor in the dish. The CKD patient is left trying to find flavor in food without using the same amounts of salt they are accustomed to or is best suited for each recipe.

The search for flavor and spices has a long standing history, dating back to the ancient Egyptians. It was responsible for great exploration and interaction with different cultures, economic growth and wealth for importers, and even the discovery and development of countries. You may recall from your history books that Mr. Columbus was searching for a faster route to India and the Spice Islands when he stumbled upon what is now the North American continent.

With all the effort made over 500 years ago to attain precious spices, it seems like a good idea to explore the flavors the ancients were looking to import.

First we will create a homemade version of garum masala, a spice mixture created from several ingredients found in India. It is often added to food after it is cooked, or as a base ingredient for other spice mixtures. You can find garum masala in stores, but putting the ingredients together can provide an understanding of the flavor profiles and allow for variations that suit individual preferences.

What we think of as curry is a mixture of the spices in garum masala with turmeric and other spices. Although we think of it as a flavor found in Indian an some Asian foods, it is believed to be an English invention, even though garum masala is a common element of the Indian kitchen.

The exact mixture of spices varies from region to region and even from household to household, but the base ingredients are very similar if not the same. This is one of the reasons I like making my own so I can control the flavor combinations.

Garum Masala (spice mixture)

Masala translates as a blend of spices, garum means "warming" or "meat", giving us a meaning of warming spices or meat spices.

When making spice mixtures I prefer to start with whole spices, when I can find them, and use my grinder to combine. Whole spices last longer on your shelf if kept properly, and produce a more intense flavor, if used soon after they are ground. If you are using powders or ground spices, be sure to keep them away from heat and sunlight. Try to use them within a year of purchase, as they lose flavor over time even when kept in a cool dry area.

Ingredients:

- 8 cardamom pods (seeds only - discard husk) or 1 tablespoon ground
- 2 tablespoons whole coriander seeds
- 1 tablespoon cumin
- 1 tablespoon whole peppercorns
- 1 teaspoon whole cloves
- 1 teaspoon fennel seed
- 1 1/2 teaspoons cinnamon
- 1/2 teaspoon ground nutmeg

Time: 5 minutes Yield: 6 Tbsp + 1 tsp Portions: 6
Serving size: 1 Tbsp

Place all whole ingredients into a spice grinder or mortar and pestle and grind thoroughly. Mix in all other ground spices and store in a sealed container, away from sunlight in a cool dry place. For most recipes I use about 1 tablespoon for 8 portions, but that can vary for each dish.

Food Note

Most pre-made spice mixtures have salt. The exact amount will not be listed, but labeling requirements dictate that ingredients be listed in the order of weight. This mean you should look for the location in the listing, or better yet, buy no salt mixtures.

Garum Masala

		Fat (g)	Sat. Fat (g)	Phos (mg)	Pot (mg)	Sodium (mg)	Calories	Carb (g)	Vit A (iu)	Vit C (mg)	Vit K (Mcg)	Vit E (mg) AT	Vit B6 (mg)
	RDA	65	20	700	3300	2300	2000	275	5000	60	120	20	2
Ingredients	Amt												
Cinnamon	1 1/2 tsp	0.00	0.00	2.50	17.70	0.40	9.50	3.00	11.45	0.15	1.20	0.10	0.00
		0.00%	0.00%	0.36%	0.54%	0.02%	0.48%	1.09%	0.23%	0.25%	1.00%	0.50%	0.00%
Cardamom	1 Tbsp	0.40	0.00	10.20	64.30	1.00	17.90	3.90	0.00	1.20	0.00	0.00	0.00
		0.62%	0.00%	1.46%	1.95%	0.04%	0.90%	1.42%	0.00%	2.00%	0.00%	0.00%	0.00%
Coriander seed	2 Tbsp	1.80	0.00	41.00	126.60	3.60	30.00	5.40	0.00	2.20	0.00	0.00	0.00
		2.77%	0.00%	5.86%	3.84%	0.16%	1.50%	1.96%	0.00%	3.67%	0.00%	0.00%	0.00%
Cumin	1 Tbsp	1.00	0.00	29.90	107.00	10.10	22.00	3.00	76.20	0.50	0.30	0.20	0.00
		1.54%	0.00%	4.27%	3.24%	0.44%	1.10%	1.09%	1.52%	0.83%	0.25%	1.00%	0.00%
Peppercorns	1 Tbsp	0.30	0.00	10.50	75.60	2.70	15.30	3.00	18.00	1.20	9.90	0.00	0.00
		0.46%	0.00%	1.50%	2.29%	0.12%	0.77%	1.09%	0.36%	2.00%	8.25%	0.00%	0.00%
Whole Cloves	1 tsp	1.30	0.40	6.60	23.87	8.60	7.00	1.33	11.47	1.77	3.07	0.20	0.00
		2.00%	2.00%	0.94%	0.72%	0.37%	0.35%	0.48%	0.23%	2.94%	2.56%	1.00%	0.00%
Fennel seed	1 tsp	0.30	0.00	9.33	32.47	1.70	6.60	1.00	2.60	0.40	0.00	0.00	0.00
		0.46%	0.00%	1.33%	0.98%	0.07%	0.33%	0.36%	0.05%	0.67%	0.00%	0.00%	0.00%
Nutmeg	1/2 tsp	0.42	0.00	2.48	4.08	0.18	6.13	0.58	1.01	0.03	0.00	0.20	0.00
Total for Dish		5.52	0.40	112.52	451.62	28.28	114.43	21.22	120.73	7.45	14.47	0.70	0.00
6 Tbsps +		8.49%	2.00%	16.07%	13.69%	1.23%	5.72%	7.72%	2.41%	12.42%	12.06%	3.50%	0.00%
Servings		12	12	12	12	12	12	12	12	12	12	12	12
Totals per serving		0.46	0.03	9.38	37.63	2.36	9.54	1.77	10.06	0.62	1.21	0.06	0.00
		0.71%	0.17%	1.34%	1.14%	0.10%	0.48%	0.64%	0.20%	1.03%	1.00%	0.29%	0.00%

Essential Amino Acid Chart Garum Masala

	Mg/ gram protein needed	Cinn	Rec. 1.5 g	Cardom	Rec. .6 g	Coriand	Reci. 1.2 g	Clove	Rec. .13 g	Fennel Seed	Rec. .3 g	Nutmeg	Rec. .13 g	Tot. g.	Tot. Am. Acid	Tot. P/g
Essential Amino Acids		p/g	Rec.	p/g	Rec.	p/g	Rec.	p/g	Rec.	p/g	Rec.	p/g	Rec.	p/g		
Histidine	18		0.00		0.00		0.00		0.00	21.11	6.33		0.00	3.86	6.33	1.64
Isoleuciine	25		0.00		0.00		0.00		0.00	44.44	13.33		0.00		13.33	3.45
Methionine	25		0.00		0.00		0.00		0.00	19.22	5.77		0.00		5.77	1.49
Leucine	55		0.00		0.00		0.00		0.00	63.67	19.10		0.00		19.10	4.95
Lysine	51		0.00		0.00		0.00		0.00	48.44	14.53		0.00		14.53	3.77
Phenylalanine	47		0.00		0.00		0.00		0.00	41.33	12.40		0.00		12.40	3.21
Threonine	27		0.00		0.00		0.00		0.00	38.44	11.53		0.00		11.53	2.99
Tryptophan	7		0.00		0.00		0.00		0.00	16.11	4.83		0.00		4.83	1.25
Valine	32		0.00		0.00		0.00			58.44	17.53		0.00		17.53	4.54
Recipe Amout		1 1/2 tsp		1 Tbsp		2 Tbps		1/2 cup		1 tsp		1/2 tsp				
Protein per recipe (g)		1.5		0.6		1.2		0.13		0.3		0.13				

Total recipe Protein (g)	# of servings	Protein / serving (g)	Not a Complete Protein
3.86	12.00	0.32	

Chicken Tikka (with rice)

Tikka means marinated pieces of vegetables or meat, traditionally cooked in a tandoor oven. A tandoor oven is made of clay or metal and is cylindrical in shape with an open top creating very high heat. The pieces of meat and veggies are skewered, then set into the oven leaning against the sides.

The sides of the oven are used to cook naan bread by flattening round pieces of dough and carefully slapping them on the insides of the clay or metal oven. If you do not happen to have a tandoor oven, not to worry. There are other ways to accomplish a delicious and very similar final dish using a standard home oven or an outdoor grill.

Please do not be reluctant to try this out because the spices are new to you or have names that do not seem familiar. This recipe is no more difficult than other marinated meat or kabob recipes that most people are more comfortable making. The spices and marinade are different, but that is all.

Ingredients:

- 2 cups white rice, uncooked
- 8 pieces chicken thighs or breasts, boneless and skin removed
- 6 ounces yogurt, full fat
- 1 cup chili paste (homemade preferred)
- 1 tablespoon garum masala
- 3 ounces fresh ginger, grated
- 3 cloves garlic, grated
- 1 tablespoon turmeric
- 1 teaspoon coriander
- 1 teaspoon cumin

Time: 4 1/2 hours Yield 24 oz. meat and 4 cups rice
Portions: 8 Serving size: 3 oz. meat plus 1/2 cup of rice.

Combine the ginger, garlic, garum masala, turmeric, cumin, and coriander and mix together thoroughly. Using a blender if possible, combine the spice mixture, yogurt, and chili paste until it is orange in color, without any white streaks.

Cut the chicken thighs into 1 inch pieces as best you can, as they do not always cooperate.

Place the meat in the yogurt and spice mixture and move to the refrigerator and marinate for 3 to 4 hours. Give the mixture a stir every hour to coat all the pieces.

About 30 minutes prior to removing chicken, soak wooden skewers in water. Place a cooling rack in a short sided sheet pan and when the meat is ready, place on skewers, covering about 2/3 of the wood.

Chicken Tikka is cooked at high heat and has small charred spots on the chicken. The high heat allows this to happen without overcooking the small pieces of meat. If you roast the skewers at a medium temperature you will not have the charred spots.

Food Note

Meat develops flavor when it is cooked, and in his 1912 paper, Louis Camille Maillard discovered that part of this flavor came from the browning on the surface of meats and other foods. What is now called the "Maillard Reaction," is the browning on the exterior of foods when cooked with dry heat. The process occurs when amino acids mix with reducing sugars at the outer surface of foods.
Salting meat for up to 30 minutes before cooking helps to bring those amino acids to the surface, increasing the reaction.
Marinating meats has the same effect, along with introducing the flavors of the marinade into the meat.

Start your rice before turning on the oven, or after you light the charcoal. It will be done at the same time as the meat.

Combine the rice and 4 cups of water in a medium sized pot. Bring to a boil, cover, and reduce to a simmer. Let sit for 15 to 20 minutes without lifting the lid.

When finished, fluff with a fork and place in a serving bowl.

Oven cooking: Set your oven to broil and place a rack in the middle of the oven. This is going to vary in every oven, but the idea is to cook the chicken through and still get the crisp outer parts. you may want to move the rack closer the broiler, depending on the space and heat in your oven.

Cook skewers for 6 to 8 minutes, rotating after 3 to 4 minutes, or when the charred spots appear. If you think the chicken has not cooked through, place the pan on a lower rack and turn off broiler. The residual heat will keep cooking the meat without further browning.

Charcoal Grilling: In an attempt to recreate the high heat of the tandoor oven, I have used both my ceramic outdoor cooker and my metal round grill. Both worked well as long as I removed the upper grate normally used for meat and vegetables. I also used longer metal skewers, but it works with well soaked wood as well.

Start the charcoal and get it nice and hot. Place the charcoal off to one side and lean the skewers point side down onto the bottom grate. The meat should be directly over the coals and leaning against the side of the grill. Cook until the underside of the meat is charred, and rotate as needed until done.

Gas Grilling: Because the heat source is in a stable position on a gas grill, cook the skewers as you would any other kabob style meat. Use high heat, but be carful not to burn the chicken.

When the meat is cooked it can be served with rice, pita, or eaten as an appetizer on it's own.

Chicken Tikka

		Fat (g)	Sat. Fat (g)	Phos (mg)	Pot (mg)	Sodium (mg)	Calories	Carb (g)	Vit A (iu)	Vit C (mg)	Vit K (Mcg)	Vit E (mg) AT	Vit B6 (mg)
	RDA	65	20	700	3300	2300	2000	275	5000	60	120	20	2
Ingredients	Amt												
White Rice, uncooked	2 cup	2.40	0.60	275.20	215.60	18.60	1350.00	296.00	0.00	0.00	0.40	0.40	0.60
		3.69%	3.00%	39.31%	6.53%	0.81%	67.50%	107.64%	0.00%	0.00%	0.33%	2.00%	30.00%
Chicken Thighs - skinless	8 Each	12.80	3.20	551.20	757.60	282.40	390.40	0.00	213.60	0.00	9.60	0.80	0.80
about 3 lbs.		19.69%	16.00%	78.74%	22.96%	12.28%	19.52%	0.00%	4.27%	0.00%	8.00%	4.00%	40.00%
Chili Paste	1 cup	6.57	0.34	136.80	1104.10	394.63	262.70	42.88	7837.28	45.60	43.44	2.70	1.52
		10.11%	1.70%	19.54%	33.46%	17.16%	13.14%	15.59%	156.75%	76.00%	3.60%	13.50%	76.00%
Yogurt - full fat	6 oz.	1.00	8.00	29.08	16.63	14.13	18.63	1.38	30.38	1.50	0.84	0.01	0.01
(1 oz. consumed)		1.54%	40.00%	4.15%	0.50%	0.61%	0.93%	0.50%	0.61%	2.50%	0.70%	0.06%	0.50%
Garum Masala	1 Tbsp	0.92	0.06	18.72	75.26	4.72	19.08	3.54	20.12	1.04	2.42	0.12	0.00
		1.42%	0.30%	2.67%	2.28%	0.21%	0.95%	1.29%	0.40%	1.73%	2.02%	0.60%	0.00%
Fresh Ginger	3 oz.	0.60	0.30	28.50	63.30	0.30	12.00	3.00	336.00	18.60	4.20	0.30	0.30
Grated		0.92%	1.50%	4.07%	1.92%	0.01%	0.60%	1.09%	6.72%	31.00%	3.50%	1.50%	15.00%
Turmeric	1 Tbsp	0.70	0.20	18.10	170.00	2.60	24.00	4.40	0.00	1.70	0.90	0.20	0.10
		1.08%	1.00%	2.59%	5.15%	0.11%	1.20%	1.60%	0.00%	2.83%	0.75%	1.00%	5.00%
Fresh Garlic	3 clove s	0.00	0.00	2.00	16.70	0.10	3.10	1.00	69.00	0.50	6.20	0.20	0.00
		0.00%	0.00%	0.29%	0.51%	0.00%	0.16%	0.36%	1.38%	0.83%	5.17%	1.00%	0.00%
Coriander	1 tsp	0.30	0.00	7.00	21.10	0.60	4.97	0.90	0.00	0.55	0.00	0.00	0.00
		0.46%	0.00%	1.00%	0.64%	0.03%	0.25%	0.33%	0.00%	0.92%	0.00%	0.00%	0.00%
Cumin	1 tsp	0.33	0.00	9.90	35.00	3.34	7.70	1.00	25.40	0.17	0.10	0.65	0.00
		0.51%	0.00%	1.41%	1.06%	0.15%	0.39%	0.36%	0.51%	0.28%	0.08%	3.25%	0.00%
Total for Dish		25.62	12.70	1076.50	2475.29	721.42	2092.57	354.10	8531.78	69.66	68.10	5.38	3.33
		39.42%	63.50%	153.79%	75.01%	31.37%	104.63%	128.76%	170.64%	116.09%	56.75%	26.91%	166.50%
# of servings		8	8	8	8	8	8	8	8	8	8	8	8
Totals per serving		3.20	1.59	134.56	309.41	90.18	261.57	44.26	1066.47	8.71	8.51	0.67	0.42
		4.93%	7.94%	19.22%	9.38%	3.92%	13.08%	16.10%	21.33%	14.51%	7.09%	3.36%	20.81%

Essential Amino Acid Chart Chicken Tikka

	Mg/ gram protein needed	White Rice	Recipe 4.2 g	Chicken Thighs Bone in	Recipe 144 g	Chili Paste	Recipe .82 g	Total Grams	Total Amino Acid	Total Protein/g
Essential Amino Acids		Protein/g	Recipe	Protein/g	Recipe	Protein/g	Recipe			
Histidine	18	23.69	99.50	31.00	4464.00	23.11	18.95	149.02	4582.45	30.75
Isoleuciine	25	43.57	183.00	52.86	7611.43	8.42	6.91		7801.33	52.35
Methionine	25	23.69	99.50	27.71	3990.86	37.56	30.80		4121.16	27.66
Leucine	55	83.57	351.00	75.00	10800.00	33.30	27.30		11178.30	75.01
Lysine	51	36.43	153.00	84.86	12219.43	22.92	18.80		12391.22	83.15
Phenylalanine	47	53.57	225.00	39.71	5718.86	26.15	21.44		5965.30	40.03
Threonine	27	36.19	152.00	42.29	6089.14	9.63	7.90		6249.04	41.93
Tryptophan	7	11.67	49.00	11.69	1682.74	30.21	24.77		1756.52	11.79
Valine	32	61.67	259.00	49.57	7138.29		0.00		7397.29	49.64
Recipe Amout		1 cup		24 oz		2 Tbps				
Protein per recipe (g)		4.2		144		0.82				

Total recipe Protein (g)	# of servings	Protein / serving (g)	Complete Protein
149.02	8.00	18.63	

Chicken Tikka Masala

This is a favorite dish for fans of Indian food, and for good reason. Another delicious recipe, with arguable origins, uses spices and cooking methods from the Indian subcontinent. Using the same process as we did for the Chicken Tikka recipe to cook the meat, we then add it to a spicy curry sauce. It can be served over rice, naan, or any other bread.

The standard sauce is made with a tomato and milk base, not always recommended for CDK patients. In this recipe I replace half of the tomato products with our homemade chili paste and use heavy cream for the dairy additive, in a lower amount.

The chili paste reduces the potassium levels found in tomatoes and the heavy cream reduces the phosphorus levels found in other milk products. The fat content, especially the saturated fat, is high and it should be a consideration for patients who are monitoring their saturated fat intake. As always, check with your care team about the levels for each portion to make sure it is a good choice for you.

Ingredients:

- 8 pieces of chicken thighs or breasts, boneless and skin removed
- 2 cups uncooked white rice
- 2 cups chili paste (homemade or low sodium)
- 6 ounces full fat yogurt
- 2 tablespoons garum masala
- 6 ounces fresh ginger, grated
- 6 cloves fresh garlic, grated or minced
- 2 tablespoons turmeric
- 2 teaspoons coriander
- 2 teaspoons cumin
- 1 tablespoon unsalted butter or ghee
- 1 cup fresh onion, chopped
- 14 ounces canned whole tomatoes
- 1 cup heavy cream
- 1 teaspoon salt
- 6 ounces water

Time: 4 1/2 hours Yield: 8 cups meat and sauce
Portions: 8 Serving size: 1 cup meat and sauce plus 1/2 cup rice

Spice mixture: The spice mixture is doubled in this recipe compared with the Chicken Tikka, with the exception of the salt. Prepare the entire mixture as in the chicken tikka recipe, and use 1/2 in the yogurt for marinating, and refrigerate the other half to use for the sauce. Use only 1/4 cup of the chili paste in the yogurt, and save the rest for the sauce.

Meat: Using the cooking method in the Chicken Tikka recipe, prepare and cook the chicken. This can be done a day ahead of time, and added to the sauce when it is prepared. I have also made this dish in one day, starting the sauce 1/2 hour before the marinated meat is ready, and cooking it in the oven while the sauce simmers.

Rice: If using, start your rice while the sauce is simmering, using the same method as the chicken tikka recipe.

Sauce: In a large dutch oven or heavy bottomed pot, on medium heat, melt the butter or ghee. Add the onions and cook for 3 to 5 minutes. Next add the remaining spice mixture and cook for 2 to 3 minutes, until the garlic is cooked but not burnt. Now add the tomatoes, chili paste, salt, and water. Simmer for 30 to 40 minutes until the sauce thickens. Be sure to break apart any tomato pieces with a wooden spoon and stir occasionally.

When the sauce has thickened, add the heavy cream and mix thoroughly. Next add the chicken pieces and simmer for 5 minutes.

Food Note

Ghee and clarified butter are cooking options in many different cuisines. They are both made from butter using medium to high heat to separate the butter into liquid fats and milk solids.

As you heat butter it will start to foam. Clear away this foam and let the butter continue to cook. The milk solids will collect at the bottom of the pan. The remaining liquid is clarified butter.

Continue cooking and the liquid starts to turn brown and develop a nutter flavor. Run this through a sieve and you end up with ghee.

Chicken Tikka Masala

		Fat (g)	Sat. Fat (g)	Phos (mg)	Pot (mg)	Sodium (mg)	Calories	Carb (g)	Vit A (iu)	Vit C (mg)	Vit K (Mcg)	Vit E (mg) AT	Vit B6 (mg)
	RDA	65	20	700	3300	2300	2000	275	5000	60	120	20	2
Ingredients	Amt												
White Rice	2 cup	2.40	0.60	275.20	215.60	18.60	1350.00	296.00	0.00	0.00	0.40	0.40	0.60
		3.69%	3.00%	39.31%	6.53%	0.81%	67.50%	107.64	0.00%	0.00%	0.33%	2.00%	30.00%
Ch. Thighs	8.00	12.80	3.20	551.20	757.60	282.40	390.40	0.00	213.60	0.00	9.60	0.80	0.80
about 3 lbs.		19.69%	16.00%	78.74%	22.96%	12.28%	19.52%	0.00%	4.27%	0.00%	8.00%	4.00%	40.00%
Chili Paste	2 cup	9.80	0.50	205.19	1647.90	1327.50	393.09	64.00	15674.60	91.20	72.34	3.69	2.65
		15.08%	2.50%	29.31%	49.94%	57.72%	19.65%	23.27%	313.49%	152.00	3.60%	18.45%	132.50
Yogurt	6 oz.	1.00	0.64	29.08	16.63	14.13	18.63	1.38	30.38	1.50	0.84	0.01	0.01
		1.54%	3.19%	4.15%	0.50%	0.61%	0.93%	0.50%	0.61%	2.50%	0.70%	0.06%	0.50%
Gar. Masala	1 Tbsp	0.92	0.06	18.72	75.26	4.72	19.08	3.54	20.12	1.04	2.42	0.12	0.00
		1.42%	0.30%	2.67%	2.28%	0.21%	0.95%	1.29%	0.40%	1.73%	2.02%	0.60%	0.00%
Ginger	3 oz.	0.60	0.30	28.50	63.30	0.30	12.00	3.00	336.00	18.60	4.20	0.30	0.30
Grated		0.92%	1.50%	4.07%	1.92%	0.01%	0.60%	1.09%	6.72%	31.00%	3.50%	1.50%	15.00%
Turmeric	1 Tbsp	0.70	0.20	18.10	170.00	2.60	24.00	4.40	0.00	1.70	0.90	0.20	0.10
		1.08%	1.00%	2.59%	5.15%	0.11%	1.20%	1.60%	0.00%	2.83%	0.75%	1.00%	5.00%
Fresh Garlic	3 clove	0.00	0.00	2.00	16.70	0.10	3.10	1.00	69.00	0.50	6.20	0.20	0.00
		0.00%	0.00%	0.29%	0.51%	0.00%	0.16%	0.36%	1.38%	0.83%	5.17%	1.00%	0.00%
Coriander	1 tsp	0.30	0.00	7.00	21.10	0.60	4.97	0.90	0.00	0.55	0.00	0.00	0.00
		0.46%	0.00%	1.00%	0.64%	0.03%	0.25%	0.33%	0.00%	0.92%	0.00%	0.00%	0.00%
Cumin	1 tsp	0.33	0.00	9.90	35.00	3.34	7.70	1.00	25.40	0.17	0.10	0.65	0.00
		0.51%	0.00%	1.41%	1.06%	0.15%	0.39%	0.36%	0.51%	0.28%	0.08%	3.25%	0.00%
Butter or Ghee	1 Tbsp	11.40	7.20	3.60	3.40	1.50	100.00	0.00	350.00	0.00	1.00	0.30	0.00
		17.54%	36.00%	0.51%	0.10%	0.07%	5.00%	0.00%	7.00%	0.00%	0.83%	1.50%	0.00%
Onion	1 cup	0.00	0.00	46.40	234.00	6.40	64.00	0.00	3.20	11.80	0.60	0.00	0.20
chopped		0.00%	0.00%	6.63%	7.09%	0.28%	3.20%	0.00%	0.06%	19.67%	0.50%	0.00%	10.00%
Tomatoes	14 oz	0.00	0.00	72.20	714.00	544.00	64.40	15.20	444.00	35.40	11.00	2.60	0.40
		0.01%	0.00%	10.31%	21.64%	23.65%	3.22%	5.53%	8.88%	59.00%	9.17%	13.00%	20.00%
H. Cream	1 cup	88.00	55.00	148.00	178.00	90.40	821.00	7.00	3499.00	1.40	7.60	2.50	0.10
		135.38	275.00	21.14%	5.39%	3.93%	41.05%	2.55%	69.98%	2.33%	6.33%	12.50%	5.00%
Salt	1 tsp					2300.00							
						100.00%							
Total for Dish		128.26	67.70	1415.09	4148.49	4596.59	3272.36	397.42	20665.30	163.86	117.20	11.77	5.16
		197.32	338.49	202.16	125.71	199.85%	163.62%	144.51	413.31%	273.09	97.67%	58.86%	258.00
8 servings		8	8	8	8	8	8	8	8	8	8	8	8
Totals per serving		16.03	8.46	176.89	518.56	574.57	409.05	49.68	2583.16	20.48	14.65	1.47	0.65
		24.66%	42.31%	25.27%	15.71%	24.98%	20.45%	18.06%	51.66%	34.14%	12.21%	7.36%	32.25%

Essential Amino Acid Chart Chicken Tikka Masala

	Mg/ gram protein needed	Rice	Rec. 16.8 g	Ch. Thigh	Rec. 144 g	Chili Paste	Rec. 7.38	Tom.	Rec. .62 g	On.	Rec. 1.8 g	H. Cream	Rec. 4.9 g	But.	Rec. .2 g	Tot. g.	Tot. Am. Acid	Tot P/g
Essential Amino Acids		P/G	Rec.	P/G	Rec.	P/G	Rec.	P/G	Rec.	P/G	Rec.	P/G	Rec.	P/G	Rec.			
Histidine	18	23.69	398.00	31.00	4464.0	23.11	170.5		0.00	12.44	22.40	27.14	133.0	27.4	5.49	17	5187.	29.53
Isoleuciine	25	43.57	732.00	52.86	7611.4	8.42	62.15		0.00	11.79	21.22	60.20	295.0	61.0	12.21		8721.	49.64
Methionine	25	23.69	398.00	27.71	3990.8	37.56	277.1		0.00	1.68	3.03	24.69	121.0	25.1	5.02		4790.	27.26
Leucine	55	83.57	1404.00	75.00	10800.	33.30	245.7		0.00	21.05	37.89	97.55	478.0	98.9	19.79		12965	73.79
Lysine	51	36.43	612.00	84.86	12219.	22.92	169.1		0.00	32.84	59.12	79.18	388.0	80.0	16.00		13447	76.54
Phenylalanine	47	53.57	900.00	39.71	5718.8	26.15	192.9		0.00	22.22	40.00	48.16	236.0	49.0	9.80		7087.	40.34
Threonine	27	36.19	608.00	42.29	6089.1	9.63	71.08		0.00	17.68	31.83	45.10	221.0	45.4	9.08		7021.	39.96
Tryptophan	7	11.67	196.00	11.69	1682.7	30.21	222.97		0.00	12.44	22.40	14.08	69.00	14.3	2.86		2193.	12.48
Valine	32	61.67	1036.00	49.57	7138.2		0.00			18.67	33.60	66.53	326.0	67.8	13.58		8533.	48.57
Recipe Amout		1 cup		24 oz		2 cups		14 oz.				1 cup		2 Tbsp				
Protein per recipe (g)		16.8		144		7.38		0.62		1.8	1	4.9		0.2				

Total recipe Protein (g)	# of servings	Protein / serving (g)	Complete Protein
175.70	8.00	21.96	

Pita Bread

Naan is the traditional bread for Chicken Tikka and Chicken Tikka Masala. It is a flat bread made with a dairy product, usually milk or yogurt, and then cooked on the side of a tandoor oven. The dairy makes traditional naan higher in phosphorus, and for that reason I prefer to use pita bread. You could certainly make naan without the dairy products, but for the most part, that is the definition of pita bread.

There are plenty of store bought pita options, however, just like store bought tortilla shells, they are high in sodium. Each pita has at least 15% of the RDI of sodium, and to keep them shelf stable, they also include other preservatives, items best avoided if your kidneys are not filtering out waste at a normal level.

For the best results cook the pita on a pizza stone or something made of cast iron. These surfaces hold heat much better than stainless steel or other metals. If you do not have either, use what you have available.

Ingredients:

- 2 1/2 cups white flour
- 1 package of yeast or 2 1/4 teaspoons
- 1 tablespoon olive oil
- 1 tablespoon sugar
- 1/2 teaspoon salt

Time: 2 hours Yield: 8 Pitas Portions: 8
Serving Size: 1 pita

Yeast is a tricky item. The process for using it depends on which type you use. The dry yeasts come in two forms, instant active and active. The difference is the size of the granules and if you use the smaller pieces, you can add the dry yeast directly to the flour mixture. When using the larger pieces, or the active dry yeast, it is important to place the yeast in warm water until it starts to foam. Remember to account for the water added to the yeast when the mixture is added to the flour.

Active yeast: Start by activating the yeast in water. Let it rest for up to 10 minutes or until it foams. If it does not foam, you will have to try again with a fresh batch. Add flour, oil, sugar, and salt into a stand mixer or a bowl if mixing by hand. Combine the ingredients thoroughly and then add the yeast mixture and water, mix until you have a soft and slightly sticky dough.

Instant active yeast: Combine all the ingredients except the water into your bowl and mix thoroughly. Add water until the dough is soft and slightly sticky.

Next step for either yeast: Let the dough mix for several minutes, or knead by hand for 6 to 8 minutes. This is an important step in any dough to help produce the glutens. Once the dough is smooth and stretchy, place it in a bowl with a 1/4 teaspoon of oil and rotate the dough so it is covered with oil. This prevents the dough from sticking to the bowl. Cover the bowl with a towel and store in a warm spot in the kitchen for about 1 1/2 hours, or until it has doubled in volume.

If you are using a pizza stone or cast iron skillet in the oven, heat them in the oven about 30 minutes before dough is ready. If you are using a cast iron or other skillet on the stove top, heat the pans after the step below.

When dough is ready, remove it from the bowl and place it on a lightly floured surface. Cut the dough into 8 equal pieces and form them into small balls. Cover and let rest for about 10 minutes. Next, roll out the each ball into 7" rounds. Be certain to add more flour to the surface if the dough is sticking to anything. Cover and let rest for another 15 minutes.

Finally, place the flattened rounds onto the heated surface for cooking. After the first minute you should see the dough puff up like a balloon. Turn the dough over and cook for another minute, or until the pita has developed some brown spots on each side. Even if some of the pieces do not puff up, turn them after a 1 1/2 minutes, and cook on the other side. Place warm pita on a plate covered with a towel until ready to serve.

Pita - White Flour

		Fat (g)	Sat. Fat (g)	Phos (mg)	Pot (mg)	Sodium (mg)	Calorie	Carb (g)	Vit A (iu)	Vit C (mg)	Vit K (Mcg)	Vit E (mg) AT	Vit B6 (mg)
	RDA	**65**	**20**	**700**	**3300**	**2300**	**2000**	**275**	**5000**	**60**	**120**	**20**	**2**
Ingredients	**Amt**												
Four, white, all purpose	**2 1/2 cups**	0.00	0.00	337.50	335.00	6.25	1137.50	237.50	6.25	0.00	1.00	0.75	0.25
		0.00%	0.00%	48.21%	10.15%	0.27%	56.88%	86.36%	0.13%	0.00%	0.83%	3.75%	12.50%
Salt	**1/2 tsp**					1150.00							
						50.00%							
Olive oil	**1 Tbsp**	13.50	1.88	0.14	0.27	0.00	119.38	0.00	0.00	0.00	8.13	1.94	0.00
		20.77%	9.38%	0.02%	0.01%	0.00%	5.97%	0.00%	0.00%	0.00%	6.77%	9.69%	0.00%
Sugar	**1 Tbsp**	0.00	0.00	0.00	0.00	0.00	48.38	12.50	0.00	0.00	0.00	0.00	0.00
		0.0%	0.0%	0.0%	0.0%	0.0%	2.4%	4.5%	0.0%	0.0%	0.0%	0.0%	0.0%
Total for Dish		13.50	1.88	337.64	335.27	1156.25	1305.25	250.00	6.25	0.00	9.13	2.69	0.25
		20.77%	9.38%	48.23%	10.16%	50.27%	65.26%	90.91%	0.13%	0.00%	7.60%	13.44%	12.50%
# of Servings		8	8	8	8	8	8	8	8	8	8	8	8
Totals per serving		1.69	0.23	42.20	41.91	144.53	163.16	31.25	0.78	0.00	1.14	0.34	0.03
		2.60%	1.17%	6.03%	1.27%	6.28%	8.16%	11.36%	0.02%	0.00%	0.95%	1.68%	1.56%

Essential Amino Acid Chart for recipe Pita

	Mg/ gram protein needed	Flour, White, AP	Recipe 32.25 g	Sugar	Recipe	Total Grams	Total Amino Acid	Total Protein/g
Essential Amino Acids		Protein/g	Recipe	Protein/g	Recipe			
Histidine	18	22.33	720.00	0.00	0.00	32.25	720.00	22.33
Isoleuciine	25	34.57	1115.00	0.00	0.00		1115.00	34.57
Methionine	25	17.75	572.50	0.00	0.00		572.50	17.75
Leucine	55	68.84	2220.00	0.00	0.00		2220.00	68.84
Lysine	51	22.09	712.50	0.00	0.00		712.50	22.09
Phenylalanine	47	50.39	1625.00	0.00	0.00		1625.00	50.39
Threonine	27	27.21	877.50	0.00	0.00		877.50	27.21
Tryptophan	7	12.33	397.50	0.00	0.00		397.50	12.33
Valine	32	40.23	1297.50	0.00	0.00		1297.50	40.23
Recipe Amout		1 1/2 cup		1 Tbsp				
Protein per recipe (g)		32.25		0.00				

Total recipe Protein (g)	# of servings	Protein / serving (g)	Not a Complete Protein
32.25	8.00	4.03	

Food groups, ingredients, and other information.

Ground Beef

Among all the food items we eat, some time should be spent addressing ground beef for a few different reasons. Beef has has been consistently at the top of the charts for meat consumption in the United States, only being surpassed by chicken in the last few years. Ground beef is also relatively inexpensive compared to other whole cuts of beef, and it is used in many different recipes. The bottom line is that we eat a lot of beef, especially ground beef, and because of this fact it is important to take a look at the different packages you find when shopping.

Unless you grind your own beef, and I strongly suggest this option if you have the inclination, time, and some sort of grinder, most people buy meat pre-ground at the market.

Each package will tell you the percentage of meat and fat as a simple ratio listed on the label. 80/20 indicates that the contents are made of 80% lean and 20% fat. 90/10 indicates that the ground beef is 90% lean and 10% fat. Lean is defined as having at least 92% beef, extra lean means it must contain at least 96% beef. This means that there is more fat than the listing on the label suggests.

Ground beef also indicates the type of cut that the meat is taken from. For instance a package labeled ground round can only be made from the lean and fat parts of the round steak. This goes for chuck and sirloin as well. Items labeled ground beef are a mix of cuts from the left over pieces of other cuts.

In *"Making Some Sense Out of Ground Beef Labeling"* (Griffin, n.d.), the guidelines used by the USDA for the term ground beef on labels are laid out and explained. By law, the meat has to come from the primal cuts and trimmings, it can contain no more than 30% fat, and can have no phosphates, liquids, binders, or other meat sources added. The primal cuts are the chuck, shoulder, rib, loin, and round, or any pieces of theses cuts known as the sub-primal.

Now that we have some definitions, let's move on to buying. The chart below identifies some of the factors to be considered for chronic kidney disease patients. As previously mentioned, and often harped upon, different patients will have different needs, and therefore should choose the product best suiting those needs.

These numbers represent ground beef, meaning a mix of cuts. You can choose to buy single cut ground beef such as ground sirloin, ground round, or ground chuck. The protein, omega-3 fatty acids, and phosphorus will all be slightly higher. Just remember that more is not always better depending on your specific needs.

Gr. Beef Lean/Fat 1 ounce	Protein (g)	Total Fat (g)	Omega 3 fatty acids (mg)	Omega 6 fatty acids (mg)	Phos. (mg)	Calories	Cholest. (mg)
80/20 - raw	4.8	5.6	13.4	122	44.2	71.1	20
- cooked	6.7	4.5	11.8	103	57.4	68.9	24.1
85/25 -raw	5.2	4.2	11.8	99.1	47.9	60.2	19
- cooked	6.9	3.9	12.6	96.3	59.1	65	24.1
90/10 - raw	5.6	2.8	9.8	76.4	51.5	49.3	18.3
- cooked	7.1	3.0	11.2	83.1	60.5	57.1	23
95/5 - raw	6.0	1.4	7.6	54	55.4	38.4	17.6
- cooked	7.2	1.7	7.8	63.9	62.2	45.9	21.3
USDA Organic Grass fed 85/25 - raw	5.4	3.6	24.6	120	49	53.8	17.4

The first column of the chart identifies the product, and the following columns represent the amounts of each heading for that product in one ounce of the ground meat. The difference in raw versus cooked is explained in the weight loss during the cooking process. The loss range varies based on the fat content and other variables. In order to have 1 ounce of cooked beef you will need to start with 1.1 to 1.2 ounces of raw (Showell, et, al., n.d.). Protein is the first column and is an important factor for both stages 1 through 4 CKD, and stage 5, ESRD. Although there is little difference in the protein content per ounce, every little bit can help, in either direction. The chart clearly shows that the protein resides in the lean parts of the beef and the higher the lean parts, the higher the protein and the lower the total fat. From a cook's standpoint, very low fat beef has less flavor, and most prefer the 80/20 cuts to bring that flavor to the final product. The common adage is "The flavor is in the fat." It just has to be the right type of fat.

The second and third column represent the amount of omega 3 and omega 6 fatty acids. There has been a great deal of research into omega 3 fatty acids in the diet of CKD patients, and in using supplementation as a dietary aid (Hu et. al. 2017) (Friedman & Moe, 2006). These studies showed varying results and suggest further research. There was some consensus that increasing omega-3 fatty acids did have a positive effect on a few of the primary problems associated with CKD. Those included cardiovascular disease, controlling blood pressure, and progression to end-stage renal disease.

Although there are other food items with significantly higher levels of omega 3 fatty acids, there is still value in optimizing the values in the food you choose. How much intake per day is not a settled science and can range from 200 mg to 1300 mg per day. Our chart shows that a 3 oz burger from USDA organic grass fed meat can supply you with 74.9 mg of omega 3 fatty acids while a 95/5 burger will supply you with only 22.8 mg. In addition the omega 6 fatty acids have positive benefits similar to omega 3's, although levels too high are not recommended (Simopoulus, 2003). This research shows that a ratio of omega 6 to omega 3 fatty acids that is too high has a negative effect on other chronic diseases. A 4:1 ratio of omega 6/omega 3 fatty acids can have a 70% reduction in the mortality rate due to cardiovascular disease, a major risk factor for CKD patients. The typical western diet has a ratio of over 15:1. The chart above shows that the USDA organic beef has a ratio of 4.9:1, while all the others are over 7.8:1.

This points out that choosing your type of ground beef should be dependent on what your medical professionals suggest is best for your particular needs. In our example above, choosing the grass fed version will give you more than 3 times the omega 3 fatty acids and a better ratio of omega 6 to omega 3's. However, depending on the lean to fat ratio, it may contain higher levels of fat. Both grass fed and very lean beef have higher levels of protein, which can be beneficial for ESRD patients, but not for

CKD patients in stages 1 through 4. As always, please discuss your particular needs with your care team.

Other issues in beef selection may be more difficult for the American consumer to navigate. Let's start from the top. Cows naturally eat grass and roam around. The grass is high is nutrients and keeps the animals healthy. The beef we eat in the U.S. is almost all factory raised beef, meaning they are fed grain, silage, and some grass. They are kept in pens and often never roam around the fields. Because the diet is not natural to cows, they are also fed antibiotics to keep them healthy, replacing the grass. These drugs also prevent loss due to other natural circumstances. The business side of beef shows that the number of cows in the US has decreased since the 1970's, but the amount of beef produced has increased, indicating that farmers have found ways to increase output from each animal, often through growth hormones ("Nation of Meat Eaters," 2012).

In European and other major beef producing countries like Argentina, Australia, and New Zealand, most of the beef is pasture raised (roaming free) and grass fed. These countries also have much stricter rules and guidelines in place for both definitions, and labeling (Olmstead, 2017).

This brings us to labeling. The U.S. has not been proactive in creating specifics for each of the qualifiers used in labeling. "Natural", "Grass Fed", and other labels have very loose guidelines, and of those, few are ever enforced. This makes the process even more difficult to manage when choosing beef. For instance "Natural" indicates that the product is all cow. Nothing else has been added to the final package. It has no requirement for how the animal was raised. "Grass Fed" means that at some point, the animal was fed grass. That's it, just at some point.

According to Olmsted (2017), the two labels that have made strides in the recent past, are "USDA Organic" and "100% grass fed." They both have a defined meaning, but enforcement is still a bit slow. These are the labels to look for if you want to avoid most of the antibiotics and hormones normally used in U.S beef.

There is no simple answer to assisting all patients with certain food choices. This is the dilemma that encompasses the chronic kidney disease diet. It applies to beef as well, so it is important to discuss the specifics with your health care team.

Chicken

The World Resources Institute (Waite, 2018) points out that chicken consumption has been rising over the years, and that the U.S. consumer currently eats more chicken than beef. A discussion on the labeling and processing can point out some of the things to look for when purchasing chicken.

There are plenty of terms used on today's labels to describe the treatment and type of chicken in the package, whether it is whole or in pieces. Just like the labeling for ground beef, some have meaning and others are merely marketing terms with no specific definitions. Outside of the general issues of eating healthy, a few of these labels can have a direct impact on the CKD patient.

One set of terms refers to raising the chickens, and the other set to processing. "Hormone free" is a term used to tell the consumer that there were no hormones added to the chicken. That not only sounds good, it is good. The problem here is that the U.S. does not allow any chicken sold to have added hormones, so this is merely restating the law.

"Antibiotic free" implies that no antibiotic medications were given to the animal or added to its feed. If the label has the USDA Organic label, these claims have an outside agency checking on them. When it comes to the "free range" or caging claims, labels are a bit deceiving. Only egg laying chickens would be kept in cages, not those used for meat. They are often housed in large warehouses like those seen on many documentaries about our food. The free range definition means the chickens have access to the outdoor areas, without any other qualifiers. It does not mean that the chicken went outside, or that it always has access, but that at some point, it could have gone outside the warehouse. Better, but still not great (Olmsed, 2017).

These are terms that have concern for flavor and treatment, but the issues for CKD patients is more pertinent in the processing. Two parts have special interest, and they have to do with sodium and phosphorus. After a chicken is processed, yes, that is what we are calling it, there are two ways to cool it down for safety purposes. One is water chilled, the other is air chilled.

Water chilled chicken is placed in a cold water solution until it reaches 39°F (4°C). Some of that solution is absorbed into the meat and it dilutes the flavor of the chicken. This is part of the "retained water" label on each package. In addition to the water from the solution, producers also inject water, called "enhancing" or "plumping", into the chicken for a variety of claimed reasons, such as maintaining moisture and adding flavor. Producers inject a salt water solution to the meat, another area of hidden sodium in the diet of the CKD patient. These producers also found out that the solution would leak out during cooking and transportation so they addd a phosphate binder to the solution to prevent this leakage.

The result of water chilled processing has now reduced flavor, which means the consumer will add other flavorings to the final product, usually salt, to better enjoy their meal. In addition producers add a sodium and phosphorus solution for a variety of reasons. Luckily this information can be found on the label, in small print and at the bottom or off to the side. Look for the lowest amount of "retained water" and low amounts for x in "Contains up to x% of water added." Both the retained water and the x should be as low as possible to limit the negative effects.

The other manner of cooling chicken is "air cooled". This allows the meat to be chilled to the appropriate temperature in an open air environment. You should still look for any added water labels to be certain of what you are buying.

Food Product 1 ounce	Protein (g)	Total Fat (g)	Omega 3 fatty acids (mg)	Omega 6 fatty acids (mg)	Phos. (mg)	Calories	Cholest. (mg)
Chicken breast with skin - raw	5.8	2.6	33.5	487	48.7	48.2	
Chicken breast with skin - roasted	8.3	2.2	30.8	395	59.9	55.2	
Chicken, dark meat with skin- raw	4.7	5.1	64.4	994	38.1	86.4	
Chicken, dark meat with skin - roasted	7.3	4.4	67.2	851	47	70.8	
Chicken breast - no skin - roasted	8.7	1	19.6	165	63.8	46.2	23.8
Chicken, dark meat- no skin - roasted	7.7	2.7	50.4	524	50.1	57.4	26

When shopping, look for chicken that is USDA organic, no antibiotics, "free" range, air chilled, and with as little water added or retained as possible. If you can't find all that, and I'm not sure most of us can, CKD patients should look for the least amount of water added and retained, that should not be too difficult to find.

The chart above shows the same information as the chart for beef. Here, there are a few factors that are interesting. In the beef section we discussed the ratio of omega 6 to omega 3 fatty acids and here you can see that ratio is even higher than all the beef listed. The ratio for chicken is between 12.6:1 to 12.8:1 when the skin is left on, and 8.4:1 for chicken breast without skin, and 10.4:1 for dark meat without skin. Once again we have different choices for different concerns for the patient.

Shrimp and other seafood

Increasing omega 3 fatty acids has been a hot topic for many years and the benefits are being studied and documented. As pointed out earlier, the ratio of omega 6 to omega 3 fatty acids has been increasing in the U.S. diet and this has been a concern based on the current research. The nutritional data has suggested that there is a benefit to having high levels of omega 6 fatty acids in the diet, provided the amount of omega 3's is also high. Promoting an increase in omega 3's is really to bring that ratio down to a healthier number.

Seafood, fish oils, and other proteins with high omega 3 fatty acids have been highly touted as a beneficial protein for this purpose. The chart below shows that the large amount of omega 3's in shrimp, and some other fish, make them a good food item on their own. In addition the very low level of omega 6 fatty acids brings down the ratio when we consider all the food we consume.

The chart shows the differences between some seafood for the variables listed. Shrimp has the highest level level of omega 3 fatty acids, and a ratio of .06:1 of omega 6 to omega 3 fatty acids when cooked. The difference in the cooking technique in the chart is based on the most common way to cook each item. Tilapia is either pan roasted, grilled, or baked, which are all dry cooking methods. As a reminder, frying is also considered a dry cooking method. Shrimp can be cooked using a dry method, or

Food Product 1 ounce	Protein (g)	Total Fat (g)	Omega 3 fatty acids (mg)	Omega 6 fatty acids (mg)	Phosphorus (mg)	Calories	Cholesterol (mg)
Shrimp - raw	5.7	0.5	151	7.8	57.4	39.7	42.6
Shrimp - cooked moist heat	5.9	0.3	97.2	5.9	38.4	27.7	54.6
Lobster - Cooked, moist heat	5.7	0.2	24.1	1.4	51.8	27.4	20.2
Cod - Atlantic - cooked, moist heat	6.4	0.2	48.2	1.7	38.6	29.4	15.4
Tilapia - cooked dry heat	7.3	0.7	67.2	84	57.1	35.8	16

boiled in water first and then eaten as is, or added to other food. I chose the moist cooking method because much of the shrimp we eat is pre cooked.

Using the ratios discussed so far, the charts show the that some fish, and shrimp especially, have positive characteristics. The significantly high levels of omega 3 fatty acids and the lower levels of omega 6 fatty acids can aid in inflammation, cardio vascular disease, and some other issues related to CKD, according to plenty of research (Hu et. al, 2017).

Another value to shrimp is the ratio of phosphorus to protein. A chart in the back of this book identifies this ratio for many foods considered high sources of protein. However, high protein comes with high phosphorus, which can be detrimental to CKD patients. The ratio used here is phosphorus to protein, and we are looking for a low ratio as well. Numbers under 7 show that phosphorus levels are lower for the levels of protein. Shrimp has a 6.5 ratio, while the 80/20 beef has an 8.5 ratio. The ratios for the different chicken is between 6.5 and 7.4.

At times these numbers may appear too small to matter, but the premise for most of this material is that small differences will add up, especially based on the few number of different meals we all eat on a recurring basis.

Any discussion about shrimp and seafood is not responsible without bringing up all the problems associated with seafood in the U.S. Labeling, sourcing, fraud, and the lack of oversight and controls, are more predominant than most people want to believe.

According to National Oceanic and Atmospheric Association (NOAA), the U.S. imported 91% of it's seafood in 2017, and of this, nearly half of it was farmed (Whittle, 2018). In addition, only one thousandth of one percent is inspected by the FDA. Most of the shrimp consumed in the U.S. is imported because the high quality of the wild caught shrimp in the U.S. is so valuable on the world market, producers can make more money selling it to other countries.

What does all this mean? A lot. Let's go through some of the research done by Larry Olmstead in his book "*Real Food, Fake Food*" (2107). There is rampant fraud in the imported seafood industry. The label may say one thing, but it may be far from what you thought you were buying. This is especially true in whitefish, as they are hard to distinguish when filleted. The sourcing of the fish is often hidden by rerouting seafood shipments to another country, relabeling the product, and then shipping it on to the U.S. This happens when authorities ban certain countries from importing seafood based on illegal chemical use and drug residues.

Kidneys are meant to filter out much of the toxins in our bodies. When they are unable to work at their best, it is extremely important to be cautious about what we are putting into our bodies. There are great health benefits to seafood, but it is important to identify the high risks as well.

When buying shrimp, look for U.S. wild caught shrimp, and only buy from a reputable location. The fraud and mislabeling goes much deeper than the retailer, and a reputable company will do its do diligence.

Studies comparing diets

Another issue that is of interest revolves around our food supply and labeling. Many areas of the world have protected specific names that can be placed on products produced in a specific region. These labels are "Designated origin of Product" labels. Certain areas of the world are known for a certain food item and the producers of the product want to protect the quality associated with the name. They create associations to determine the guidelines and oversee the quality and compliance. These food associations work with the government to enforce the restrictions on the name, area, and production format. This maintains the high quality of the product, and protects the brand from low quality substitutes that can dilute the name and product.

Countries have their own specialty products and through trade agreements protect each others rights. They do this by restricting the use of designated products when not associated with the specific region and producers. This is true in many countries around the world, except the U.S.

The easiest example is Parmigiano Reggiano. Only certain cows, eating certain grass, from a certain region, can produce the milk. The cheese can only be molded in certain containers, in one size, and must be aged for a certain amount of time. No other ingredient other than the particular milk, salt, and rennet can be used in making the cheese. It is then inspected by the associations to be certified for quality. That is why it is so good.

This is the same process for many other products around the world. Olmstead (2017) points out many of these, and the reasoning behind the structure.

In the U.S. we do not have such a system or protection for foods from other countries. We allow producers to place nearly any label on a product and sell it as the real thing. We essentially make our own rules.

Why is this important? In any study comparing the diets and health of other countries, it is important to note that we are not comparing apples to apples when it comes to food. It's more like we are comparing apples to rotten oranges. Anyone who has eaten pesto from the Ligurian region in Italy will tell you that they never had pesto until that moment. What they had was some odd green concoction made to look, and appear to taste, like pesto.

There is a Latin phrase "Ceteris Paribus," and it means "all other things being equal." In many of the studies that compare diets from other parts of the world to the U.S. diet, the premise is that the same food in each region is equal and can be an equal replacement, however, this is not the case.

I do not mean to disparage the professionals who conduct these important studies, but merely to point out the problem with one of the premises in hopes that it can be accounted for.

The differences in food discussed above, including the processing, types of feed, regulatory guidelines, and derivation of the product are important factors that should be part of these studies for the benefit of us all.

An example of these studies is the mediterranean diet vs. the American diet. Researchers compare the diet and health of people living in each area. Coronary heart disease, various types of cancer and other diseases are significantly lower in the mediterranean than in the U.S. (Dontas, et. al., 2007). Diet and environmental issues are targeted as the cause, including exercise. Professional researchers certainly understand the effect of the environment on the individual, but the environment and processing guidelines for the food each group consumes may be overlooked or too difficult to account for.

My position does not contradict the results of the research of those mediterranean diet studies, but suggests it is not just the choice of foods we eat, but also the foods that our foods eat, and the rules of processing that are governed by each area.

Parmigiano Reggiano

Parmigiano Reggiano is a hard cheese made in a very specific region in Italy, and overseen by both the producers and the government. There are very specific rules about the milk used, the cows that produce the milk, the food that can be fed to the cows, the size of the forms used to make the cheese, and on and on and on.

The name is protected in Italy, the European Union, and several other countries in the world. No additives are allowed, no antibiotics can be fed to the cows, and only three ingredients can be used to produce the cheese. Each wheel is inspected throughout the aging process, and must pass each test to meet quality standards.

It is a controlled product through and through. The quality of real Parmigiano Reggiano is quite noticeable, and wonderful. This is also true for many other cheeses produced is certain areas of the world. Roquefort must be aged in certain caves in the French town Roquefort. In those caves lives a unique mold spore, and a constant temperature and humidity, all of which are part of the quality of the cheese. The milk comes from specific sheep, eating certain grass, grown in certain areas.

What is the point of all of this and why does it matter to the CKD patient? Hard cheeses are allowed in moderation in the diet of many patients. It is lower in phosphorus because much of the moisture is lost through aging. That moisture is the whey, and it contains a higher level of protein, and hence phosphorus. It is usually eaten is smaller amounts and that adds to the limited phosphorus.

All sounds good so far. Now the problem. You most likely are not eating Parmigiano Reggiano. Those little green or red cans you buy in the store are "Parmesan" or "Parmesan Topping." They are usually filled with cheddar, or some other type of cheese product along with other flavorings and preservatives. The taste is nothing like the high quality stuff, and most will use even more of it to get the flavor out the product.

These and other pre-grated cheeses will also add cellulose to the product as an anti-clumping agent. You may have heard the hoopla about tree bark in these cheeses, but cellulose is a broad category of plant fiber that is not considered harmful. It does dilute the flavor of whatever it is mixed with, and you may use more of the product, or salt, in order to get the desired flavor. The real problem these reports were pointing out is the percentage of cellulose that was found in the mix. The law allows for up to 4% cellulose, but many contained between 8% - 12%. More labeling and monitoring problems.

So where does this leave us when buying cheese? I suggest always buying your cheese in block form, and be sure you can see part of the words Parmigiano Reggiano on the rind. The real stuff has this imprinted on the forms used for the cheese. If it does, it is imported from the proper region in Italy and you can be certain of the ingredients.

The next step is grating the cheese for use. I only grate when I am ready to serve or finish a dish. Letting it sit around after grating exposes too much air to the cheese and this will dilute the flavor. This goes for shaving the cheese or serving it in small pieces as well.

Now for the proper tool. There are plenty of graters out in the world with different size holes. Each produces a different end product that can be useful for different purposes. Shaving a thin piece is perfect for a platter of hors d'oeuvres of cheeses. Box graters have different sizes for mixing with other cheeses, or on top of veggies or pizza.

But my favorite is a zester. A zester creates very small and light pieces that look fluffy when piled up. A microplane is a long, thin, and flat zester attached to a handle. When used, the same volume of the cheese compared to other graters has a significantly less weight. The pre-grated store bought cheese weighs 4 times as much per volume than cheese grated on a zester. This is why many of my recipes have a lower amount of phosphorus and sodium listed in the charts.

The photo shows 1 ounce of Parmigiano Reggiano cheese using a microplane on the left and on the right is 1 ounce of store bought pre-grated. The volume measure is 1 cup for the microplane and 1/4 cup for the store bought pre-grated.

Absorption

In each section of this book we discuss the nutrients in food and different methods to maximize or minimize them, depending on each individual's needs. The body requires a variety of nutrients to sustain itself and has different ways to access them. Some nutrients can be produced by the body, others require external input.

A simple example is the essential amino acids. Of the 20 amino acids, 9 are considered "essential", meaning the body cannot produce them and they must come from outside sources. The other amino acids can be produced by the body. Vitamins, on the other hand, all come from outside sources. The body does not produce them and we get them primarily from food. One exception is vitamin D which we get from sunlight or supplements. Very few foods have vitamin D, and those that do, have it added and are labeled "Fortified with Vitamin D."

Supplements are another outside source of nutrients and can increase the levels of certain vitamins and minerals. They are a big business and often misused or misunderstood, and a doctor can help you better understand what may or may not be helpful. But remember, if a healthy body is getting too much, it rids itself of the excess, provided the input amounts do not exceed the pace at which it is processed out. Those of us with CKD do not easily handle overloads or deficiencies of nutrients and this can cause serious problems.

Within some of our food you can find different types of nutrients, organic and inorganic. Organic nutrients are those that are found in the food, before any processing and additives. Fresh fruits and vegetables, grains, meats, beans, and nuts, are just some examples. However, when processing foods for large scale production or longer shelf life, additives are used and these are the inorganic parts. They include phosphates, potassium, nitrates, sugar, sodium, or any combination of these.

In the CKD community it is important to understand the difference between the organic and inorganic nutrients in food products. Bioavailability is one of the main reasons. Phosphorus found naturally is absorbed at 40%-60% in meat products, and a little less in vegetable based foods. However, added phosphorus in foods is absorbed at 100% (Noori, et. al.2010). CKD patients have problems removing the excesses in our bodies, and this particular one can have very detrimental effects.

Vitamins are broken down into 2 categories, water soluble and fat soluble. Vitamins A, D, E, and K, are fat soluble, meaning that they need fat to be absorbed into the body. They are best eaten with foods that have some fat content. Once they are absorbed, these vitamins can be stored in the liver and the body's fat, and then distributed as needed. Because of this you do not need daily doses in your diet.

Water soluble vitamins are the B's and C. They absorb in the body easily but are not stored, meaning the vitamins need to be ingested daily for proper health.

Vitamin D is a slightly different story. As mentioned above, very few foods contain vitamin D and the primary sources are sunlight and supplements. Milk and some other dairy products are often fortified with it, but dairy can cause phosphorus issues for CKD patients.

Vitamin D from sunlight does not need fat to be absorbed into the body. Supplements and foods, either naturally or fortified, do require a moderate amount of fat for the best results. It is an important vitamin in the diet of CKD patients because it contributes to the absorption of calcium providing for stronger bones.

Fiber is another important factor in food even though it is considered a non nutrient because it does not absorb into the body. Fiber is either soluble or insoluble and both have an impact on the body. Soluble fiber grabs onto the bad cholesterol and stops it from absorbing in the body. Insoluble fiber helps with constipation, which can be an issue for those on dialysis and if you are taking certain medications.

Food preparation and cooking methods can also impact the bioavailability of nutrients in our diet. Almost all food loses some of its nutrients when cooked. Even the cooking method, dry heat vs moist heat, will impact the amount of loss in each item. When cooking veggies in water, nearly half of the water soluble vitamins and will leach into the water, along with smaller amounts of minerals. This is not always a negative issue for cooking methods.

Chopping and cooking vegetables increases the bioavailability of their nutrients, especially those with rigid tissue structures. Cooking also increases the digestibility of foods and can rid them of harmful bacteria. This is helpful for those who become immunocompromised after a transplant.

You can reduce high phosphorus levels by using a moist heat cooking method like braising, boiling, or stewing, and use as little of the cooking liquid as possible (Ando, et. al., 2015). The liquid absorbs both the positive and the negative nutrients for the CKD patient, but if your phosphorus levels are high it is a good cooking method for proteins to reduce some of the phosphorus.

Chopping up potatoes and soaking them in room temperature water helps leach out the potassium. Boiling also increases potassium loss. This food item that CKD patients are generally told to avoid can benefit from this process and cooking method.

One exception are the carotenoids. These are red, yellow, and orange fat soluble pigments founds in foods like tomatoes, carrots, and others. Carotenoids have a variety of potential health benefits including conversion into vitamin A, and increasing other nutritional components. Cooking tomatoes and carrots increase the bioavailability of these carotenoids in the body (Ostrenga, 2018).

All this information adds to an already difficult nutritional regiment for the CKD patient, but each small application can make a difference to the quality of life by increasing general health. The valuable information here is a deeper understanding of why doctors and nutritionists recommend fresh fruit and vegetables, limited processed foods, and supplements, to provide a balance in the body that was once being done by the kidneys.

Whole Wheat Flour and other Grains

One of the first dietary restrictions suggested for CKD patients concerns whole grains. Breads, rice, or anything else made using whole grains are higher in potassium and phosphorus and need to be controlled. As kidney function starts to decline so does the ability to filter out these and other minerals. Having high levels of potassium and phosphorus in the body can cause additional damage to our bodies.

It is important to understand exactly what the term "whole wheat" or "whole grain" means before we start to look into this potential dietary restriction. Most bread is made from flour that comes from wheat. Wheat is a type of grass, and it is cultivated for the seed or kernel, which is a cereal grain. These are the little pieces that look a little like rice at the top of the tall grassy plant.

Those seeds have three parts, bran, germ, and endosperm. The endosperm makes up 83% of the total weight of the kernel, and is used to make standard white flour. This is done by removing the bran and the germ. Whole wheat flour is made using all three parts of the wheat kernel. The bran makes up 14.5% of the kernel by weight and the germ about 2.5%. I know this is getting very "science-y" but we will use this information soon.

The bran and germ contain plenty of nutrients which are lost when using only the endosperm for white flour. The term "enriched" means that some of these nutrients are added back into the flour. The term "fortified" means nutrients that are not normally present are added to the end product.

After the endosperm of the wheat kernel is milled, the flour is aged for about 3 months. During this time, the flour turns from a pale yellow to a whiter shade. It also develops proteins and glutens. "Bleached" is just that, a process to make the flour look as white as possible, without waiting the 3 months. "Bromated" means that potassium bromate was added to enhance the bread making ability of flour. With all this going on to make white flour faster, whiter, longer lasting on shelves, and then to appear as healthy as it was before all his started, is it any wonder that bread has been demonized, and associated with new health issues.

Many of these terms also apply to other grains, so be sure to read those labels. One of the changes in labeling is to list white flour in bread as wheat flour, as it comes from wheat rather than another grain. That's the good part. If the label says whole wheat flour, then the flour uses all three parts of the kernel. Wheat flour only uses the endosperm. It is both a bit deceiving and informative, as many breads now use flour from several other types of grain. This three part construction is the same for other grains including rice, rye, and barley. White rice is just brown rice with the bran and germ removed. Rye flours are made by using different amounts of the bran and germ with the endosperm. The darker the rye flour, the more of the bran and germ that is being used. Other grains like barley, quinoa, bulgar, etc, use all three parts and are considered whole grains.

The reason restrictions were placed on whole grains and products made using whole grains is the higher levels of potassium and phosphorus in the bran and germ. This applies to both wheat, rice, and most others grains.

The two charts below shows the increase in both minerals when the bran and germ are used for rice and flour made from all three parts.

Flour (100 g)	Endosperm	Bran	Germ	Total Bran and Germ	Total Phosphours	Total Absorbable Phosphorus
White (wheat) flour	100%	0%	0%	0%	108.00	108.00
Phosphorus (mg)	108	0	0		108.00	108.00
Whole wheat flour	83%	14.5%	2.5%	17%	357.00	89.64
Phosphorus (mg)	89.64			267.36	357.00	89.64

A study by Williams et. al. (2013) is now showing that the existence of the additional phosphorus in the bran and germ are bound to organic compounds that do not allow for the release and absorption of the phosphorus. Phylate is the main compound that is bound to the phosphorus. It requires a separate outside compound phylase to release it and allow for absorption. There is very little phylase in the bran and germ, and it is reduced during milling, processing, and over time. Some organizations are now pointing this out, however, it is slow to move through the system and into the dietary programs for patients.

The charts also show the difference in the phosphorus amounts when using white versus whole wheat flours, or white versus brown rice, and compares the absorption based on this research.

100 g	Phosphorus (mg)	Potassium (mg)	Serving Size	Phosphorus per serving (mg)	Potassium per serving (mg)
White Flour - all purpose	108	107	28 grams or 1 ounce	30	30
Whole Wheat Flour	357	363	28 grams or 1 ounce	100	102
White Rice - cooked (about 1/2 cup)	37	29	1/2 cup or 125 grams 4.4 ounces	37	27
Brown Rice - cooked (about 1/2 cup)	77	79	1/2 cup or 125 grams 4.4 ounces	77	77

The charts show the total phosphorus in whole wheat flour and brown rice is higher, however the amount of flour created by the endosperm is lower. If phosphorus in the bran and germ in whole wheat flour is not absorbed, and the phosphorus from the endosperm is lower, then the total phosphorus absorbed is lower using whole wheat flour.

Some grains do not have any phylase at all, and this indicates that they may not be as high in absorbable phosphorus as originally proposed. These include oatmeal, millet, and corn (Williams, et. al. 2013).

The increase in potassium is less of an issue based on the total amount allowed in the diet. The recommended daily allowance is 3300 mg, and the increase to under 80 mg is not very large.

Sugar, Diabetes, and Dietary Fats

Chronic kidney disease is a risk factor for diabetics. An article on the American Diabetes Association website states that up to 23% of diabetics have chronic kidney disease, and 45% of patients on dialysis have kidney failure caused by diabetes (Cavanaugh,

2007). There is also a high risk of diabetes after kidney transplantation called new-onset diabetes after transplant, or NODAT. Although there is not a great deal of data or consensus on the incidence rate, some have shown that the rate is up to 30% or higher. They do agree that the medications needed after transplant can interfere with the body's ability to properly process sugar.

Sugar is a disaccharide, a combination of two monosaccharides. It is a chemical bond that when separated, is the basis of energy in the body. The common monosaccharides are glucose, fructose, and galactose. Different combinations of these create the disaccharides. Sucrose is combination of glucose and fructose, lactose is glucose combined with galactose, and maltose is glucose connected to another glucose. Many foods will contain these disaccharides naturally but we also add them to processed food to add desired sweetness, as a preservative, or to balance out bitter flavors.

Our bodies break down the food we eat and use it for a variety of purposes, including energy and providing nutrients. The energy comes from the carbohydrates and they are found in fruits, vegetables, milk, and grains. These are called complex carbohydrates. In order for the body to access the energy it must break down the food into single units of monosaccharides. Complex carbohydrates also have valuable nutrients the body needs.

Simple carbohydrates include monosaccharides, and disaccharides and when we add these to food items, we do not get the benefit of the nutrients that come from the complex carbohydrates.

Once the body breaks down the food into a monosaccharide, and it enters the blood stream, the hormone insulin is released into the blood stream to assist in getting the energy into the cells of the body. The pancreas creates insulin, and helps regulate the glucose levels in the blood. When the amounts of insulin produced are insufficient, glucose builds up in the blood and can cause a variety of health problems including blindness, heart disease, lower limb amputation, and kidney disease.

Here are some facts and studies that I find interesting on this subject matter that apply to our discussions about food and its makeup.

In the 1960's through the early 1970's, researchers were looking for evidence for dietary consumption relating to coronary heart disease (CHD). The two supposed culprits were saturated fats and sugar. Early researchers suggested refined carbohydrates, especially sugar, was the culprit, along with a low intake of dietary fiber. Simply put, eating too much added sugar and less fiber was causing increased risk of CHD. Others proposed that saturated fatty acids (SFA) were the culprits and they needed to be restricted (Temple, 2018).

The research battle concluded with the generally accepted guideline that SFA's were to blame and should be restricted. This went on for nearly 50 years, and in 2014 the prevailing opinion began to change. Research from as early as 1991, along with more recent data, showed that the argument for saturated fatty acids was exaggerated and pointed out that the sugar industry paid for the research implicating SFA's, and researchers never disclosed the funding (Kearns, et. al., 2016).

There is now growing research that suggests that refined carbohydrates, including sugar, have an increased role in coronary heart disease and that saturated fatty acids play a much smaller role (Temple, 2018).

During that same time period, a Pew Research Center article (Desilver, 2016) using U.S.D.A. data shows the changes in the American diet since 1970. One notable factor was the peak of sugar and corn sweeteners in 1999 at 90.2 pounds per year. In 1970 the total was around 68 pounds per year, however that number has been reduced to 77.3 pounds per year in 2014.

Additional research has shown that our energy intake has increased since the 1970's and this increase has been primarily from carbohydrates. Those carbohydrates are found in processed and frozen foods, and take out meals (Chun et. al., 2010).

Let's keep going with some more research. It has been found that more than 75% of the energy intake in the American diet comes from processed foods. The data was taken from the years 2000 to 2014. This includes minimally processed to ultra processed. The study also indicates that all processed food is higher in saturated fats, sugar, sodium, and other non natural flavorings (Poti, et. al., 2015).

And finally, according the CDC's report "Long Term Trends in Diabetes (n.d), the percentage of patients with diabetes has increased from .93% in 1958 to 7.4% in 2015.

If we look at this research together it suggests the saturated fatty acids, and not sugars, were targeted as a primary cause of coronary heart disease in the 1960's, and should be restricted in our diets. Twenty five years later sugar consumption increased over 30% in our diets, mostly from processed foods. New research started making the rounds and sugar consumption was reduced to an increase of about 13% in 2016 from the 1970 level.

Our energy intake increased during the same time period along with our caloric intake, and 75% of our total calories was now coming from some form of processed food, which is full of sugar, fats, sodium, and non natural flavors.

It may be difficult to know if the increase in the consumption of processed food and the associated added sugars are responsible for the diabetes increase. It may be difficult to determine if this is correlation or causation, especially in healthy individuals. However, the risk factors for chronic kidney disease patients to be diagnosed with diabetes, and for those with diabetes to have chronic kidney disease has been proven. Additionally, those added sugars are known risks for both groups of patients.

I have provided this information to stress the importance of diet as a mechanism of control for our own health. This applies to everyone, but groups of us with certain health issues are encouraged to be much more conscious of the food we are eating.

The recipes in this book include certain core ingredients that we use on their own and in other dishes. Using this process to reduce the dependency and intake of processed food is another argument for going through the effort, along with the better taste and controlled sodium levels.

Phosphorus and Protein

Protein is an important part of any diet, and by diet I mean the food we eat, not an attempt to lose weight. We all need protein to help build and repair muscles, skin, and fight off infections. It also helps build hormones and enzymes the body needs to survive. Proteins are comprised of amino acids, which are called the building blocks of life for these exact reasons.

A functioning kidney filters the blood of excessive nutrients, fluid, and waste. They allow the rest to remain in the blood stream to supply the body with what it needs, including proteins. As your kidney function declines, some of the protein that is normally returned to the blood stream passes through and ends up in your urine. This can also further damage the kidneys.

When the body breaks down protein for use, it creates protein waste. This waste is filtered out by healthy kidneys, but can return to the blood stream as your kidneys start to decrease in function. If you reach end-stage renal disease (ESRD) and are placed on dialysis, the machines remove the waste that the kidneys did not, and in the process also remove some of the protein that we need.

For these reasons, protein needs change as the disease progresses. Doctors will recommend decreasing your protein intake in the early stages of kidney failure to slow the progression of the disease. High protein diets can cause increased pressure on the glomeruli (the filter parts in the kidney), and cause further damage (Ko et. al., 2017).

Protein intake is increased at ESRD as dialysis removes both the protein waste and the protein needed to build and repair the body, including the healing of needle sticks and fighting off potential infections from both hemodialysis and peritoneal dialysis.

Phosphorus is primarily responsible for the health of bones and teeth. It does have other functions but this one gets the most focus. Phosphorus is the second most abundant mineral in the body after calcium. It is also found it many foods we eat and is especially high in foods rich in protein like meat, seafood, dairy, and beans.

Healthy kidneys filter out excess phosphorus from the blood to maintain the proper proportions the body needs. Unhealthy kidneys do not. I am hoping medical professionals will excuse the following explanation as it is oversimplified, but I think it might help the average patient like myself. Phosphorus is strongly associated with calcium and the body tries to keep a balance of the two in the blood. If the phosphorus levels increase in the blood, calcium needs to be balanced out and this can happen by taking it out of the bones, hence weakening them.

This process can place too much calcium in the blood and the body, forming deposits that cause even more damage. Once again, this is overly simplified, and leaves out some of the reactions in the body including the parathyroid and the intestines. This may ring a bell for dialysis patients who suffer from an increase in the parathyroid hormone (PTH) and have to add medication to maintain proper levels.

This explanation is merely meant to increase the understanding for patients and their families, and provide some explanation for the importance of the charts that follow. They provide information about different foods considered high sources of protein. They also show the associated amounts of fats and phosphorus along with portion sizes and, if needed, how the food is served, and the amount for that serving. As an example, a single chicken drumstick will provide about 1.5 ounces of meat, but a serving size is 3.0 ounces. There are others in the list that are similar, but most are straight forward, 3 ounces of ground pork is just that, and it is one serving.

Fat and saturated fat content are listed, as proteins from meat have significant amounts of both. It is important to take notice of these numbers, especially if fat needs to be controlled in your diet.

The highlighted column is the ratio of phosphorus to protein. That is the amount of phosphorus in the serving size divided by the amount of protein. The lower the ratio means that there is a lower level of phosphorus per gram of protein. This can be positive if you need to increase your protein, and control your phosphorus. Keep in mind, however, that a small ratio could still indicate a level of phosphorus that is not recommended for all patients.

As mentioned above, at certain levels of kidney failure you may have different protein needs, so be sure to discuss your needs with your health care professionals.

I have also included 2 different cooking methods for many of the food items, where appropriate. One is a dry cooking method, usually roasting. The second is a wet cooking method, braising or boiling. Some foods are rarely if ever boiled or braised and therefore do not have both methods listed.

The importance of the cooking method is shown in the phosphorus levels. It has the same effect as boiling vegetables. Some of the nutrients from the vegetable will transfer to the water, and braising has similar effects on meat, moving some of the phosphorus to the cooking liquid. This can help reduce the levels of phosphorus, provided you do not use all the left over liquid for the recipe. Chefs will tell you that the remaining liquid is like gold for most dishes as it contains concentrated flavors from the cooking process and can intensify the taste. This is another example of the difficulties in creating meals for CKD patients that are also tasty.

Seafood, Crustaceans, and Shellfish - Phosphorus to Protein Ratio

Name	Serving Size	Unit 1	Serving Size	Unit 2	Phos. (mg)	Protein (g)	Ratio (g/mg)	Fat (g)	Sat. Fat (g)	Food unit	Amt. oz	Unit 1	Amt. g	Unit 2
Crab, Blue, cooked, moist heat	3	oounce	85	grams	243	23.8	10.21	1.5	0.2	1 crab	4.5	oz..	127	g
Crab, Alaskan King, cooked, most heat	3	ounces	85	grams	186	15.5	12.00	1.3	0.1	1 leg	4.8	oz.	134	g
Lobster, cooked, moist heat	3	ounces	85	grams	157	17.4	9.02	0.5	0.1	1 tail	4.0	oz.	113	g
Clams, cooked, moist heat	3	ounces	85	grams	287	21.7	13.23	1.7	0.2	10 small	3.0	oz	85	g
Clams raw	3	ounces	85	grams	144	10.9	13.21	0.8	0.1	6 medium	3.0	oz	85	g
Mussels, cooked, moist heat	3	ounces	85	grams	242	20.2	11.98	3.6	0.7	10 small	3.0	oz.	85	g
Shrimp, cooked, moist heat	3	ounces	85	grams	116	17.8	6.52	0.9	0.2	8-9 medium	4.0	oz	112	g
Oysters, raw	3	ouines	85	grams	79.1	4.4	17.98	1.3	0.4	6 medium	3.0	oz	85	g
Salmon, dry heat	3	ounces	85	grams	204	17.4	11.72	10.5	2.1	1/2 filet	7.1	oz.	198	g
Tilapia, cooked, dry heat	3	ounces	85	grams	171.3	21.9	7.82	2.1	0.9	1 filet	4.0	oz	85	g
Tuna steak, cooked,dry heat	3	ounces	85	grams	277	25.4	10.91	5.3	1.4	1 steak	4.3	oz	120	g
Tuna, light ,canned in water	3	ounces	85	grams	139	21.7	6.41	0.7	0.2	2/3 of a can	3.0	oz	85	g
Catfish, domestic, cooked dry heat	3	ounces	85	grams	208	15.9	13.08	2.4	0.6	1 filet	5.1	oz	143	g
Cod, Atlantic, dry heat	3	ounces	85	grams	190	19.5	9.74	0.7	0.1	1 fillet	3.2	oz	90	g
Haddock, cooked, dry heat	3	ounces	85	grams	205	20.6	9.95	0.8	0.1	1 fillet	5.4	oz.	`190	g
Sardines, canned in oil	3.75	ounces	92	grams	451	22.7	19.87	10.5	1.4	1 can	3.8	oz	92	g
Octopus, cooked, moist heat	3	ounces	85	grams	237	25.3	9.37	1.8	0.4	3 ounces	3.0	oz	85	g
Squid, raw	3	ounces	85	grams	188	13.2	14.24	1.2	0.3	3 ounces	3.0	oz	85	g
Squid, squid	3	ounces	85	grams	213	15.2	14.01	6.4	1.6	3 ounces	3.0	oz	85	g

Meat: Beef, Poultry, Pork, Lamb - Phosphorus to Protein Ratio

Name	Serving Size	Unit 1	Serving Size	Unit 2	Phos. (mg)	Protein (g)	Phos/Protein Ratio (g/mg)	Fat (g)	Sat. Fat (g)	Food unit	Amt.	Unit 1	Amt.	Unit 2
Chicken, Drumsticks, meat & skin Roasted	3	oz.	85	grams	147	22.8	6.45	9.3	2.7	1 drumstick,	1.86	oz.	52	g
- Stewed	3	oz.	85	grams	118.5	21.3	5.56	9	2.4	1 drumstick,	1.86	oz.	52	g
Chicken, Thighs, meat & skin Roasted	3	oz.	85	grams	146.1	21	6.96	12.9	3.6	1 Thigh, Boneless	2.21	oz.	62	g
- Stewed	3	oz.	85	grams	116.7	19.5	5.98	12.3	3.6	1 Thigh, Boneless	2.21	oz	62	g
Chicken, Breasts, meat & skin Roasted	3	oz.	85	grams	179.7	24.9	7.22	6.6	1.8	1/2 Breast	3.50	oz.	98	g
- Stewed	3	oz.	85	grams	131.1	23.1	5.68	6.3	2.1	1/2 Breast	3.50	oz.	98	g
Chicken, Wings, meat & skin, Roasted	3	oz	85	grams	126.9	22.5	5.64	6.6	1.9	1 wing,	1.21	oz	34	g
Chicken, Wings, meat & skin, fried	3	oz	85	grams	126	21.9	5.75	7.1	1.9	1 wing	1.42	oz	32	g
Chicken, Drumsticks, meat only Roasted	3	oz	85	grams	154.5	23.7	6.52	4.8	1.2	1 let, boneless	1.57	oz	44	g
- Stewed	3	oz	85	grams	126	23.1	5.45	4.8	1.2	1 let,	1.57	oz	44	h
Chicken, Thighs, meat only, Roasted	3	oz	85	grams	153.6	21.9	7.01	9	2.4	1 Thigh, Boneless	1.86	oz	52	g
- Stewed	3	oz	85	grams	125.1	21	5.96	8.1	2.4	1 Thigh, Boneless	1.86	oz	52	h
Chicken, Breasts, meat only, Roasted	3	oz	85	grams	189	26.1	7.24	3	0.9	1/2 Breast	3.07	oz	86	g
- Stewed	3	oz	85	grams	138.6	24.3	5.70	2.4	0.6	1/2 Breast	3.07	oz	86	g
Chicken, Wing, meat only, Roasted	3	oz	85	grams	139.5	25.5	5.47	3.3	0.9	1 wing	0.75	oz	21	g
Bacon, cured, strips, cooked	1	oz	28	grams	157	10.7	14.67	10.3	3.6	3 slices	1.00	oz.	28	g
Bacon, cured, Canadian style, grilled	1	oz	28	grams	82.9	6.8	12.19	2.4	0.8	2 slices	1.68	oz	47	g
Ham, cured, boneless, Roasted	3	oz	85	grams	239	19.2	12.45	2.5	0.9	3 oz	3.00	oz	85	g
Pork Belly, Raw	2	oz	53	grams	60.4	5.2	11.62	29.6	10.8	2 oz piece	2.00	oz	56	g
Pork, Back ribs, Roasted	3	oz	85	grams	186	24.3	7.65	25.1	9.3	3-4 Ribs	3.00	oz	85	g
Pork Tenderloin, Roasted	3	oz	85	grams	265.5	18.3	14.51	12.5	4.6	1 slice	3.00	oz	85	g
Pork Loin, Roasted	3	oz	85	grams	255.3	20.4	12.51	20.9	7.8	1 slice	3.00	oz	85	g

Name	Serving Size	Unit 1	Serving Size	Unit 2	Phos. (mg)	Protein (g)	Phos/Protein Ratio (g/mg)	Fat (g)	Sat. Fat (g)	Food unit	Amt.	Unit 1	Amt.	Unit 2
Ground Pork, cooked	3	oz	85	grams	192	21.8	8.81	17.7	6.6	1 patty	3.00	oz	85	g
Pork Shoulder, Picnic Ham, Braised	3	oz	85	grams	180	23.8	7.56	10.4	3.5	3 oz shredded	3.00	oz	85	g
Pork Shoulder, Picnic Ham, Roasted	3	oz	85	grams	180	19.8	9.09	10.7	3.7					
Italian Sausage	4	oz	83	grams	141	15.9	8.87	22.7	7.8	1 link	4.00	oz	83	g
Pickled pig's feet	2	oz	56	grams	48.8	6.6	7.39	2.8	0.8	1 pigs foot	2.00	oz	56	g
Spareribs	3	oz	85	grams	222	24.7	8.99	26	9	5 ribs	4.00	oz	112	g
Turkey, Breast, Roasted	3	oz	85	grams	181.5	24.3	7.47	2.7	0.6	1/2 breast	12.10	oz	344	g
Turkey, Leg, Roasted	3	oz	85	grams	168	24	7.00	4.5	1.5	1 leg (no bone)	8.60	oz	245	g
Turkey, Dark meat, Roasted	3	oz	85	grams	164.7	24.3	6.78	9.6	3	3 oz torn	3.00	oz	85	g
Ground Turkey	3	oz	85	grams	164.7	23.1	7.13	11.1	2.7	1 patty raw	4.00	oz	113	g
Lamb, leg, sirloin, Braised	3	oz	85	grams	141	24.1	5.85	11.4	4.8	1 piece (1	5.29	oz	148	g
Lamb, leg, sirloin, Roasted	3	oz	85	grams	157	21.2	7.41	16.7	7	1 piece (1 lb raw)	9.29	oz	260	g
Lamb, leg, shank, Braised														
Lamb, leg, shank, Roasted	3	oz	85	grams	170	22.7	7.49	9.7	3.9	1 piece	9.54	oz	267	g
Lamb, ground	3	oz	85	grams	171	21	8.14	16.7	6.9	1 patty (raw	4.00	oz	113	g
Sirloin Steak, Broiled	3	oz	85	grams	184	23.1	7.97	13.4	5.3	1 steak	11.80	oz	306	g
Brisket, Cured, Dry heat (corned Beef)	3	oz	85	grams	106	15.4	6.88	16.1	5.4	3 oz				
Brisket, Braised	3	oz	85	grams	143.7	24	5.99	14.7	5.7	3 oz	3.00	oz	85	g
Chuck, Roasted	3	oz	85	grams	178	20.9	8.52	10.4	3.6	3 oz	3.00	oz	85	g
Chuck, Braised	3	oz	85	grams	156	25.6	6.09	15.7	6.2	3 oz	3.00	oz	85	g
Flank, Roasted	3	oz	85	grams	177.3	23.4	7.58	6	2.4	1 steak	13.54	oz	379	g
Flank, Braised	3	oz	85	grams	227	23.8	9.54	6	2.4	3 oz	3.00	oz		
Ground, 80 % lean, Pan broiled	3	oz	85	grams	174	20.4	8.53	13.5	5.1	1 Patty	3.00	oz	85	g
Rib, Roasted	3	oz	85	grams	147	19.6	7.50	24.2	9.8	1 piece (1 lb raw)	10.43	oz	292	g
Round, Roasted	3	oz	85	grams	135	22.1	6.11	10.6	4	1 piece (1 lb raw)	12.07	oz	338	g

Name	Serving Size	Unit 1	Serving Size	Unit 2	Phos. (mg)	Protein (g)	Phos/Protein Ratio (g/mg)	Fat (g)	Sat. Fat (g)	Food unit	Amt.	Unit 1	Amt.	Unit 2
Round, Braised	3	oz	85	grams	173	27.8	6.22	10.1	3.8	1 piece (1 lb raw)	10.04	oz	281	g
Porterhouse, Roasted	3	oz	85	grams	159	19.8	8.03	18.8	7.2	1 piece (1 lb raw)	9.11	oz	255	g
Tenderloin, Roasted	3	oz	85	grams	178	21.5	8.28	18.9	7.5	1 steak, raw	5.50	oz	154	g
T-Bone, Roasted	3	oz	85	grams	162	20.4	7.94	17.3	6.7	1 piece (1 lb raw)	9.25	oz	259	g
Beef, grass fed, strip steaks,, lean only, raw	3	oz	85	grams	178.2	19.5	9.14	2.4	0.9	1 steak	7.70	oz	214	g
Beef, strip steaks, lean only, raw	3	oz	85	grams	177.3	18.6	9.53	3	1.2	1 steak	7.70	oz	214	g
Beef, grass fed, ground, raw	3	oz	85	grams	147	16.2	9.07	10.6	4.5	1 patty	3.00	oz	85	g
Beef, ground, raw	3	oz	85	grams	132.6	14.4	9.21	16.8	6.3	1 patty	3.00	oz	85	g

Nuts/Beans - Phosphorus to Protein Ratio

Name	Serving Size	Unit 1	Serving Size	Unit 2	Phos. (mg)	Protein (g)	Phos/Prot. Ratio (g/mg)	Fat	Sat. Fat (g)	Food unit	Amt.	Unit 1	Amt.	Unit 2
Almonds, Blanched	1	oz.	28	grams	134	6.1	21.97	14.2	1.1	23 Almonds	1.0	ounces	28	g
Cashews, raw	1	oz.	28	grams	185	5.1	36.27	12.3	2.2	18 medium	1.0	oz.	28	g
Chestnuts	1	oz.	28	grams	30	0.9	33.33	0.4	0.1	3 chestnuts	1.0	oz.	28	g
Coconut, meat, raw	1	oz.	28	grams	31.6	0.9	35.11	9.4	8.3	1/3 cup	1.0	oz.	28	g
Hazelnuts	1	oz.	28	grams	81.2	4.2	19.33	17.5	1.3	18 nuts	1.0	oz.	28	g
Macadamia nuts	1	oz.	28	grams	53.1	2.2	24.14	21.3	3.3	12 nuts	1.0	oz.	28	g
Pecans	1	oz.	28	grams	77.5	2.6	29.81	20.8	1.8	15 halves	1.0	oz.	28	g
Pistashios	1	oz.	28	grams	137	5.8	23.62	12.9	1.6	49 nuts	1.0	oz.	28	g
Walnuts	1	oz.	28	grams	144	6.7	21.49	18.3	1.7	14 halves	1.0	oz.	28	g
BEANS (dried or canned)														
Black beans, boiled	1	cup	172	grams	241	15.2	15.86	0.9	0.2	1 cup	6.0	oz	172	g
Fava, boiled	1	cup	126	grams	212	12.9	16.43	0.7	0.1	1 cup	4.5	oz	126	g
French, boiled	1	cup	177	grams	161	12.5	12.88	1.3	0.1	1 cup	6.0	oz	177	g
Great Northern, boiled	1	cup	177	grams	292	14.7	19.86	0.8	0.2	1 cup	6.0	oz	177	g
Kidney, boiled	1	cup	177	grams	251	15.3	16.41	0.9	0.1	1 cup	6.0	oz	177	g
Navy, boiled	1	cup	182	grams	262	15	17.47	1.1	0.2	1 cup	6.5	oz	182	g
Pinto, boiled	1	cup	171	grams	251	15.4	16.30	1.1	0.2	1 cup	6.0	oz	171	g
Green Snap beans	1	cup	156	grams	31.2	1.80	17.33	0.3	0.1	1 cup	5.6	oz	156	g
Lima, immature, (green)	1	cup	170	grams	221	11.5	19.22	0.5	0.1	1 cup	6.0	oz	170	g
Lima, mature (white)	1	cup	188	grams	209	14.7	14.22	0.7	0.2	1 cup	6.7	oz	188	g
Peas , green	1	cup	170	grams	113.8	7.6	14.97	0.4	0.1	1 cup	6.0	oz	170	g
Peanut, Roasted with Salt	1.5	oz.	42	grams	150	9.9	15.15	20.8	4.4	2/3 cup or 28 nuts	1.5	oz	42	g
Peanut Butter	2	Tbsp	32	grams	115	8	14.38	16.1	3.4	2 Tbsp	1.1	oz	32	g
Sugar snap peas whole	1	cup	63	grams	33.4	1.8	18.56	0.1	0	10 pods	1.2	ounces	34	g
Beans, Fresh														
Green, snap	1	cup	126	grams	36.2	2.4	15.08			1 cup	4.5	oz	126	g
Peas (sugar snap or snow)	1	cup	98	grams	51.9	2.7	19.22			1 cup	3.5	oz	98	g

Name	Serving Size	Unit 1	Serving Size	Unit 2	Phos. (mg)	Protein (g)	Phos/Prot. RatiO (g/mg)	Fat (g)	Sat. Fat (g)	Food unit	Amt. oz	Unit 1	Amt.
Milk, 2%	8	oz	244	grams	229	8.1	28.27	4.7	2.9	1 cup	8.0	oz	244.00
Milk, whole (3.25% milk fat)	8	oz	244	grams	222	7.9	28.10	7.9	4.6	1 cup	8.0	oz	244.00
Half & Half	8	oz	242	grams	230	7.2	31.94	27.8	17.3	2 Tbps	1.0	oz	28.00
Light Whipping Cream	8	oz	239	grams	232	5.2	44.62	73.9	46.2	2 Tbps	1.0	oz	28.00
Heavy Whipping Cream	8	oz	148	grams	148	4.9	30.20	88.1	54.8	2 tbsp	1.0	oz	28.00
Butter, no salt	0.5	oz	14	grams	3.4	0.1	34.00	11.4	7.2	1 Tbsp	0.5	oz	14.00
Margarine, 60% fat	0.5	oz	14	grams	2.3	0		8.6	1.6	1 Tbsp	0.5	oz	14.00
Lard	0.5	oz	13	grams	0	0		12.6	5	1 Tbsp	0.5	oz	13.00
Yogurt, plain, skim	8	oz	245	grams	385	14	27.50	0.4	0.3	1 cup	8.0	oz	245.00
Yogurt, plain, low fat	8	oz	245	grams	353	12.9	27.36	3.8	2.5	1 cup	8.0	oz	245.00
Yogurt, plain, whole milk	8	oz	245	grams	233	8.5	27.41	8	5.1	1 cup	8.0	oz	245.00
CHEESE													
American	1.6	oz	44	g	216	9.4	22.98	10.6	6.1	2 slices	1.6	oz	44.80
Swiss	2	oz	56	g	318	15	21.20	15.6	10	2 slices	2.0	oz	56.00
Cheddar	2	oz	56	g	286	14	20.43	18.6	10.8	2 slices	2.0	oz	56.00
Provolone	2	oz	56	g	278	14.4	19.31	15	9.6	2 slices	2.0	oz	56.00
Goat cheese	2	oz	56	g	143.4	10.4	13.79	16.8	11.6	4 Tbsp	2.0	oz	56.00
Ricotta, part skim	2	oz	56	g	102.4	6.4	16.00	4.4	2.8	4 Tbsp	2.0	oz	56.00
Ricotta, whole milk	2	oz	56	g	88.4	6.4	13.81	7.2	4.6	4Tbsp	2.0	oz	56.00
Queso Fresco	2	oz	56	g	124	12	10.33	16.8	10.6	4 tbsp	2.0	oz	56.00
Feta	2	oz	56	g	188.8	8	23.60	12	8.2	4 tbsp	2.0	oz	56.00
Cottage	2	oz	56	g	91.2	6.6	13.82	2.9	0.9	1/2 cup	4.0	oz	112.00
Cream Cheese	2	oz	56	g	59.4	3.4	17.47	19.2	10.8	4 Tbsp	2.0	oz	56.00
Mozzarella, part skim, low M	2	oz	56	g	230	12	19.17	11.2	7	4 Tbsp	2.0	oz	56.00
Brie	2	oz	56	g	105.2	11.6	9.07	15.4	9.8	4 Tbsp	2.0	oz	56.00
Fontna	2	oz	56	g	193.8	14.2	13.65	17.4	10.8	2 Slices	2.0	oz	56.00
Danish Blue	2	oz	56	g	216	12	18.00	16	10.4	4 Tbsp	2.0	oz	56.00
Gorgonzola	2	oz	56	g									
Gouda	2	oz	56	g	306	14	21.86	15.4	10.8	4 Tbsp	2.0	oz	56.00
Parmigiano-Reggiano	2	oz	56	g	394	21.2	18.58	14.4	9.2	4 Tbsp	2.0	oz	56.00
Pecorino Romano	2	oz	56	g	426	17.8	23.93	15	9.6	4 Tbsp	2.0	oz	56.00

References

Blood tests for transplant. (2018, July 12). Retrieved from https://www.kidney.org/atoz/content/BloodTests-for-Transplant

Chronic kidney disease - Diagnosis and treatment - Mayo Clinic. (2018, March 8). Retrieved from https://www.mayoclinic.org/diseases-conditions/chronic-kidney-disease/diagnosis-treatment/drc-20354527

Compound Found In Peppers Can Drastically Impact The Health Of Those With Chronic Kidney Disease. (2013, November 29). Retrieved from https://www.kidneybuzz.com/compound-found-in-peppers-can-drastically-impact-the-health-of-those-with-chronic-kidney-disease/2013/11/29/compound-found-in-peppers-can-drastically-impact-the-health-of-those-with-chronic-kidney-disease

Fluid Overload in a Dialysis Patient. (2018, March 30). Retrieved from https://www.kidney.org/atoz/content/fluid-overload-dialysis-patient

Kidney basics. (n.d.). Retrieved from http://kidney.org

Peritoneal Dialysis | NIDDK. (2018, January 1). Retrieved from https://www.niddk.nih.gov/health-information/kidney-disease/kidney-failure/peritoneal-dialysis

Nutrient Data Laboratory (U.S.), & Consumer and Food Economics Institute (U.S.). (1999). USDA nutrient database for standard reference. Riverdale, Md: USDA, Nutrient Data Laboratory, Agricultural Research Service

Retrieved from https://www.cdc.gov/diabetes/statistics/slides/long_term_trends.pdf

Stages of Chronic Kidney Disease. (n.d.). Retrieved from https://www.davita.com/education/kidney-disease/stages

Sugar Substitutes for Diabetes | Joslin Diabetes Center. (2019, February 25). Retrieved from https://www.joslin.org/info/the_abcs_of_sugar_substitutes.html

The factors that modify glycemic indexes | Official web site of the Montignac Method. (n.d.). Retrieved from http://www.montignac.com/en/the-factors-that-modify-glycemic-indexes/
Types of Fat. (2018, July 24). Retrieved from https://www.hsph.harvard.edu/nutritionsource/what-s should-you-eat/fats-and-cholesterol/types-of-fat/
Vascular access for hemodialysis. (n.d.). Retrieved from https://surgery.ucsf.edu/conditions--procedures/vascular-access-for-hemodialysis.aspx
Vitamin A. Fact sheet for health professionals. (n.d.). Retrieved from *https://ods.od.nih.gov/factsheets/VitaminA-HealthProfessional/
Ando, S., Sakuma, M., Morimoto, Y., & Arai, H. (2015, November 25). The Effect of Various Boiling Conditions on Reduction of Phosphorus and Protein in Meat. - PubMed - NCBI. Retrieved from https://www.ncbi.nlm.nih.gov/pubmed/26163744
Aubrey, A., & Charles, D. (2012, June 27). A Nation Of Meat Eaters: See How It All Adds Up. Retrieved from https://www.npr.org/sections/thesalt/2012/06/27/155527365/visualizing-a-nation-of-meat-eaters
Cavanaugh, K. L. (2007, July 1). Diabetes Management Issues for Patients With Chronic Kidney Disease. Retrieved from http://clinical.diabetesjournals.org/content/25/3/90
Chen, O. K., Chung, C. E., Wang, Y., Padgitt, A., & Song, W. O. (2010, August 3). Changes in Intakes of Total and Added Sugar and their Contribution to Energy Intake in the U.S. Retrieved from https://www.ncbi.nlm.nih.gov/pmc/articles/PMC3257707/
Cicetti, F. (2012, July 13). What's the Most Common Blood Type? Retrieved from https://www.livescience.com/36559-common-blood-type-donation.html
Deliver, D. (2016, December 13). How America's diet has changed over time. Retrieved from http://www.pewresearch.org/fact-tank/2016/12/13/whats-on-your-table-how-americas-diet-has-changed-over-the-decades/

Dontas, A. S., Zerefos, N. S., Panagiotakos, D. B., & Valis, D. A. (2007, March). Mediterranean diet and prevention of coronary heart disease in the elderly. Retrieved from https://www.ncbi.nlm.nih.gov/pmc/articles/PMC2684076/

Eby, G. A., & Eby, K. L. (2006, March 20). Rapid recovery from major depression using magnesium treatment. - PubMed - NCBI. Retrieved from https://www.ncbi.nlm.nih.gov/pubmed/16542786

Friedman, A., & Moe, S. (2006, March 1). Review of the Effects of Omega-3 Supplementation in Dialysis Patients. Retrieved from https://cjasn.asnjournals.org/content/1/2/182

Griffen, D. (n.d.). Making Some Sense Out of Ground Beef Labeling - Meat Science. Retrieved from https://meat.tamu.edu/ground-beef-labeling/

Hayek, T., Ito, Y., Verdery, R. B., Aalto-Setälä, K., Walsh, A., & Breslow, J. L. (n.d.). Dietary fat increases high density lipoprotein (HDL) levels both by increasing the transport rates and decreasing the fractional catabolic rates of HDL cholesterol ester and apolipoprotein (Apo) A-I. Presentation of a new animal model and mechanistic studies in human Apo A-I transgenic and control mice. Retrieved from https://www.ncbi.nlm.nih.gov/pmc/articles/PMC288145/

Hu, J., Liu, Z., & Zhang, H. (2017, January). Omega-3 fatty acid supplementation as an adjunctive therapy in the treatment of chronic kidney disease: a meta-analysis. Retrieved from https://www.ncbi.nlm.nih.gov/pmc/articles/PMC5251198/

Johnston, C., & Gaas, C. (2006, May 30). Vinegar: Medicinal Uses and Antiglycemic Effect. Retrieved from https://www.ncbi.nlm.nih.gov/pmc/articles/PMC1785201/

Jolayemi AT, A. T., & Ojewole, J. A. (2013, June 13). Comparative anti-inflammatory properties of Capsaicin and ethyl-aAcetate extract of Capsicum frutescens linn [Solanaceae] in rats. - PubMed - NCBI. Retrieved from https://www.ncbi.nlm.nih.gov/pubmed/24235936

Kearns, C. E., Schmidt, L. A., & Glantz, S. A. (2016, November 1). Sugar Industry and Coronary Heart Disease Research. Retrieved from https://jamanetwork.com/journals/jamainternalmedicine/article-abstract/2548255

Ko, G. J., Obi, Y., Tortorici, A. R., & Kalantar-Zadeh, K. (2017, January 20). Dietary protein intake and chronic kidney disease. - PubMed - NCBI. Retrieved from https://www.ncbi.nlm.nih.gov/pubmed/27801685

Machoy-Mokrzynska, A. (n.d.). Fluoride-magnesium interaction. Retrieved from www.mgwater.com/fl2.shtml

McGee, H. (2003). On Food and Cooking: The Science and Lore of the Kitchen. New York, NY: Scribner.

Mosley, M. (2014, October 16). Is reheated pasta less fattening?. Retrieved from http://www.bbc.com/news/magazine-29629761

A Nation Of Meat Eaters: See How It All Adds Up. (2012, June 27). Retrieved from https://www.npr.org/sections/thesalt/2012/06/27/155527365/visualizing-a-nation-of-meat-eaters

Noori, N., Simms, J. J., Koople, J. D., Shah, A., Coleman, S., Shinaberger, C. S., ... Bross, R. (2010, April 4). Organic and inorganic dietary phosphorus and its management in chronic kidney disease. - PubMed - NCBI. Retrieved from https://www.ncbi.nlm.nih.gov/pubmed/20404416

Olmsted, L. (2017). Real Food/Fake Food: Why You Don't Know What You're Eating and What You Can Do About It. Chapel Hill, NC: Algonquin Books.

Ostrenga, S. (2018, January 30). Are you absorbing the nutrients you eat?. Retrieved from https://www.canr.msu.edu/news/are_you_absorbing_the_nutrients_you_eat

Rippe, J. A., & Angelopoulos, T. J. (n.d.). Relationship between Added Sugars Consumption and Chronic Disease Risk Factors: Current Understanding. Retrieved from https://www.ncbi.nlm.nih.gov/pmc/articles/PMC5133084/

Rohrig, B. (n.d.). Hot Peppers: Muy Caliente - American Chemical Society. Retrieved from https://www.acs.org/content/acs/en/education/resources/highschool/chemmatters/past-issues/archive-2013-2014/peppers.html

Showell, B. A., Williams, J. R., Duvall, M., Howe, J., Patterson, K., Roseland, J. M., & Holden, J. M. (n.d.). USDA Table of Cooking Yields for Meat and Poultry. Retrieved from https://www.ars.usda.gov/ARSUserFiles/80400525/data/retn/usda_cookingyields_meatpoultry.pdf

Simopoulos, A. P. (2002, October). The importance of the ratio of omega-6/omega-3 essential fatty acids. - PubMed - NCBI. Retrieved from https://www.ncbi.nlm.nih.gov/pubmed/12442909

Spillett, R. (2014, July 24). How 60% of people each the same seven regular meals every week. Retrieved from https://www.dailymail.co.uk/news/article-2703772/What-s-dinner-tonight-Lasagne-just-like-week-How-60-people-seven-regular-meals-everyweek.html

Temple, N. J. (2018, January 4). Fat, Sugar, Whole Grains and Heart Disease: 50 Years of Confusion. Retrieved from https://www.ncbi.nlm.nih.gov/pmc/articles/PMC5793267/

Waite, R. (2018, January 20). 2018 Will See High Meat Consumption in the U.S., but the American Diet is Shifting | World Resources Institute. Retrieved from https://www.wri.org/blog/2018/01/2018-will-see-high-meat-consumption-us-american-diet-shifting

Weisenberger, J. (2013, September). Heart-Healthy Fats — It's the Type—Not the Amount—That Matters. Retrieved from http://www.todaysdietitian.com/newarchives/090313p14.shtml

Wentzel, U. O., Herbert, L. A., Stahl, R. A., & Krenz, I. (2006, March 1). My Doctor Said I Should Drink a Lot! Recommendations for Fluid Intake in Patients with Chronic Kidney Disease. Retrieved from http://cjasn.asnjournals.org/content/1/2/344.full

Whittle, P. (2018, June 24). US imported more seafood in 2017 than any prior year. Retrieved from https://

www.apnews.com/aac21ba31ed346069d2a8105faf6fa1f
Williams, C., Rocco, C., & Kotnako, P. (2013, December 20). Whole grains in the renal diet--is it time to reevaluate their role? - PubMed - NCBI. Retrieved from https://www.ncbi.nlm.nih.gov/pubmed/24496192
Wolke, R. L. (2010). Tools and technology. In What Einstein Told His Cook: Kitchen Science Explained. New York, NY: W. W. Norton & Company.
Zheng, J., Zheng, S., Feng, Q., Zhang, Q., & Xiao, X. (2017). Dietary capsaicin and its anti-obesity potency: from mechanism to clinical implications. Bioscience reports, 37(3), BSR20170286. doi:10.1042/BSR20170286. (30). Retrieved from https://www.ncbi.nlm.nih.gov/pmc/articles/PMC5426284/

Index

CPSIA information can be obtained
at www.ICGtesting.com
Printed in the USA
BVHW021626170220
572578BV00001B/2